THE ECONOMICS OF INNOVATION IN THE TELECOMMUNICATIONS INDUSTRY

John R. McNamara

Q

Quorum Books_____

New York • Westport, Connecticut • London

Library of Congress Cataloging-in-Publication Data

McNamara, John R.
    The economics of innovation in the telecommunications industry /
John R. McNamara.
       p. cm.
   Includes bibliographical references and index.
   ISBN 0–89930–558–X (alk. paper)
    1. Telecommunication—United States—Technological innovations.
I. Title.
   HE7775.M38 1991
   384′.041—dc20   91–15932

British Library Cataloguing in Publication Data is available.

Library of Congress Catalog Card Number: 91–15932
ISBN: 0–89930–558–X

First published in 1991

Quorum Books, One Madison Avenue, New York, NY 10010
An imprint of Greenwood Publishing Group, Inc.

Printed in the United States of America

The paper used in this book complies with the
Permanent Paper Standard issued by the National
Information Standards Organization (Z39.48–1984).

10 9 8 7 6 5 4 3 2 1

Contents

Preface

The telecommunications infrastructure of the United States is of great economic significance in its own right as well as being essential for the efficient functioning of all other sectors of the economy. An efficient telecommunications system not only makes possible the rapid provision of government services, the instantaneous transmission of news and the simplification of social communications but also is essential for the widespread communication of prices and other economic data that form the basis for the countless business decisions that, in the aggregate, guide the behavior of the free-market economy. Improvements in telecommunications system efficiency and services directly improve the functioning of market mechanisms and therefore have a direct impact on the performance of the economic system. There is no reason to accept anything but the best available telecommunications services, and the national interest is well served by active research and development programs that bring improved communications services to market.

Since the Civil War, when the emerging telegraph service proved invaluable in the conduct of military operations, an efficient, reliable, and secure national electronic communications system has been considered strategically important by the federal government. The government also has an interest in the availability of electronic communications to assist in protecting the public in time of natural disaster or epidemic.

Early in the history of telecommunications both in Europe and in the United States, it was a matter of common sense that a region or a nation was best served by a single integrated electronic communications system that could connect the sender of a message with a recipient. The assumption underlying this belief was the technological fact that the parties must be connected by a physical wire path. Since any two parties would be in actual communication for only a small fraction of the time, why not let all customers share common lines and switching facilities to the greatest extent possible? It seemed obvious that the construction and operation of a wire communications network could be accomplished most efficiently by a single organization. Such an organization, due to its exclusive control of electronic communications, would have monopoly power and could take economic advantage of its customers.

For these reasons, and because electronic communications are also useful for criminal purposes and extralegal political activities and are therefore potentially dangerous to existing governments, European nations established government monopolies over telephone and telegraph systems, often as adjuncts to postal services. Such institutional arrangements are still common in Europe today.

In the United States a different path was chosen. Congress decided to permit private ownership of telecommunications companies but to place them under the control of governmental regulatory agencies. Such agencies, often assisted by the courts, were to monitor the behavior of the telecommunications firms, regulate their activities and adjudicate disputes arising from their actions. Telecommunications firms, in return for agreeing to their status as regulated public utilities, were both guaranteed a fair return on their investments and protected from competitive entry into their markets.

In the beginning, these arrangements appeared reasonable and practical, and they eventually led to the establishment of the immensely successful American Telegraph and Telephone (AT&T) Company, arguably the most efficient telecommunications company in the world. The rigid standardization imposed by AT&T no doubt assisted the rapid growth of telephone service throughout the nation.

The national telecommunications monopoly was founded on the technological premise that electronic communications require physical wire paths connecting all customers. Over the years a continued belief in this premise came to be essential in order to justify the telecommunications monopoly, just as the same premise is being used today to justify local cable television monopolies. Radio communications, coming into commercial development around the turn of the century, were a threat to the premise. The balkanization of radio technology through the dispersion of crucial patents among many firms held up the commercialization of AM radio, a pattern that was repeated each time a new radio technology, for example, FM, became available. It was the intervention of government, stressing military necessity, in both world wars that accelerated the development of most of the advanced radio technologies in commercial use today. The perceived shortage of radio frequencies in the preferred parts of the spectrum and the implication that transmission capacity would therefore be limited relative to the essentially unlimited capacity of land lines also played a role in the delay of innovations in radio communication.

A subsidiary reason for the reluctance to abandon hard-wire electronic communications was the regulatory principle of fair rate of return on assets. Hard-wire networks necessarily involve a great deal of assets, and those assets have very long lives, earning a fair return throughout. Why abandon such a comfortable business for the uncertainties of advanced technologies? Regulators and the public came to believe in the inherent goodness of the national telecommunications monopoly with its predictability, rigid rules and monotonously reliable performance, all captured by the term then in common use to describe the company: "Ma Bell."

The breakup of AT&T in 1984, the eventual result of technologically based attacks in court by competitors, has released a flood of pent-up communications advances, some of which are still restricted in the United States by court-imposed limitations while being actively pursued overseas.

The solid-state revolution in electronics combined with advances in computer technology have greatly expanded the capacity limitations of radio communications while reducing system investment costs. It is essential that regulatory officials promptly recognize and respond to developing technological opportunities in telecommunications, ideally by deregulating telecommunications markets where practicable or by altering regulatory policy to encourage efficiency and innovation. For example, the allocation of radio frequencies will continue to be a responsibility of the FCC, but the allocation procedure used should result in the most efficient use of the spectrum.

The critical importance of advanced telecommunications to the efficient functioning of the economy and to the well-being of the public requires continuing attention to the evolution of government regulation of the industry as new telecommunications technologies are developed. The nation deserves no less.

This book summarizes the history of telecommunications in the United States, emphasizing the interaction of government regulation, business decisions, market structure and technological advances in the industry. The economic arguments supporting continuation of telecommunications monopoly on a national or regional basis are considered, and the economic consequences of advancing technology are shown to counter these arguments. The telecommunications technologies of the recent past, the present and the emerging technologies of the future are described, and their economic implications are assessed. The effects of regulatory practice on the rate of technological innovation in telecommunications are studied and compared with the likely behavior of unregulated markets in which free rein is given to innovation.

Leaving the comfortable world of "Ma Bell" for the uncertainties and temporary inconveniences of new business arrangements and new technologies has occasionally proven to be unsettling to telecommunications customers and awkward for government officials. But, as one takes the long view, choosing to provide the freedom to innovate and compete is the way the United States achieved greatness, and in this writer's opinion the preservation and expansion of such freedom will provide at least as many benefits in the future as it has in the past.

Acknowledgments

The ideas for this book arose out of a long-standing interest in the regulation of public utilities and the effects of such regulation on innovation. This interest was reinforced by attendance at the AT&T-sponsored program on the Economics of Regulated Public Utilities at the Graduate School of Business, University of Chicago in 1976, followed by the completion of series of research projects and papers on the economics of public utilities. Gerald Faulhaber, while visiting Lehigh University to present a paper, also stimulated my interest in this topic. Joseph Szmania of AT&T suggested that I write a paper on innovation in telecommunications for a special edition of the *Review of Business*, and the resulting article "Pricing and Technological Change in Regional Telecommunications Companies" appeared in that journal in 1989. Liliane Miller, then associated with Quorum Books, proposed a book-length treatment of innovation in the telecommunications industry based on the ideas contained in the above article, resulting in present book. The editors at Quorum Books, especially Eric Valentine and Andrew Schub, have been very helpful in bringing this project to a successful conclusion, and I thank them.

Dean Richard W. Barsness of the College of Business and Economics, Lehigh University provided summer support and other assistance and encouragement for this work. My colleagues Robert Thornton, Chair Department of Economics, and Eli Schwartz generously provided ideas, research materials and counsel, and I am grateful to them.

My wife, Nancy, and our children, Tom, John and Amy, have been especially supportive through the period of preparation of this book and their cheerful encouragement has been helpful as well as making the writing task pleasant.

THE ECONOMICS OF INNOVATION IN THE TELECOMMUNICATIONS INDUSTRY

CHAPTER 1 _____

Introduction

Beginning with competing patent applications for the invention of the telephone in 1876, the history of telecommunications in the United States is a history of technological advances closely followed by legal maneuvering intended to secure exclusive business rights. Alexander Graham Bell's application for a patent on the telephone was filed only hours before Elisha Gray's application. The award of the patent to Bell and his financial backers rather than to Western Union, the purchaser of Gray's rights, played a large role in determining the future course of the telecommunications industry in the United States. The Bell organization evolved into the American Telephone and Telegraph Company (AT&T), the leading telecommunications company in the world for over 100 years.

The management of the Bell System never lost sight of the importance of technological advances as both a source of new business opportunities and protection from competition through the patent laws. The Bell Laboratories subsidiary of AT&T was a source of unparalleled technological advances over the years and is arguably responsible for the commanding position in the world occupied by the U.S. telecommunications industry today, but Bell Labs also built the technological foundation for a legal wall around the company's businesses.

By 1980 the telecommunications sector of the U.S economy exceeded the size of the agricultural sector. The common carriers in the telecommunications industry were earning annual revenues in excess of $60 billion, and were on their way to revenues exceeding $100 billion by the end of the decade. Investment in new telecommunications plant and equipment accounted for more than one-twelfth of gross fixed capital formation in the United States. Before its recent breakup, AT&T was the largest private corporation in the world, its 1 million employee work force exceeded only by that of the U.S. government. The former Bell System was widely admired for its business success and for its traditions of public service and technological excellence.

However, an inevitable conflict arises when a business enterprise grows so large and so powerful that it begins to substitute its own judgment, no matter how excellent, for that of the citizens of a great democracy. By the 1970s, as technological advances in telecommunications began to accelerate around the world, spurred on by AT&T's development of the transistor and readiness to make solid-state technology available to all, it was becoming clear to knowledgeable observers that AT&T's monopoly of telecommunications equipment and services was no longer in the best interests of the American public, and the economic power of AT&T, whether it was being used wisely or not, was the source of growing resentment.

The telecommunications revolution is profoundly changing the way we live and do business. Technological advances are rendering the immense existing investment in copper-wire lines and conventional telephone switching equipment in the United States obsolete. The former monolithic AT&T system has been replaced by a new industry structure, still not completely formed. The appropriate postures of federal and state regulatory agencies towards the telecommunications industry are still to be determined. The move away from analog signals transmitted over land lines towards the digital switching of digital signals over radio frequencies has profound implications for the mix of products and services that can be provided, for the cost structure of the industry, and for the appropriate form of government regulation.

An understanding of the process by which technological advances have brought about profound change in an industry that successfully withstood change and preserved its regulated monopoly status for most of a century is important in its own right, but such an investigation also reveals insights concerning the technological policy of government in a democracy when the public welfare is vitally affected by the performance of key high-technology industries. What role will telecommunications play in the U.S. economy in coming years? What role does government regulation play in promoting technological advances? What should government policy be concerning the telecommunications industry?

The menu of technological opportunities from which business chooses consists of the body of existing inventions, sometimes available through

R&D activities, sometimes available through licensing or other agreements with inventors or patent holders, and sometimes available through legal challenges or imitation. Innovations are those inventions selected for commercial development because their economic prospects are bright. While the process of invention occurs in many places for many reasons, the innovation process is influenced by the economic climate and market structure and by government policies and regulation. Economic advantages can be gained through R&D activities both by converting inventions into profitable innovations and by patenting inventions with the sole intention of denying them to competitors and, incidentally, denying their advantages to consumers. Thus, the legal system plays a major role in the innovation process, but it is a double-edged sword.

The basic telephone patent was granted to Alexander Graham Bell in 1876, and in the next few years the nucleus of the Bell System was formed to exploit the Bell patent. From the very beginning, the leaders of the industry desired total control of the market. Relevant technological advances were sought out and patents were purchased, for use in the Bell system when advantageous, to keep the patents out of the hands of potential competitors when not immediately useful.

By 1900 some 900 patents were held by the Bell System, including the vitally important patent on the loading coil, a simple device whose utility was governed by sophisticated scientific theory. The loading coil permitted efficient long distance telephone service. The exclusive control over long distance service provided a means for interconnecting affiliated local telephone systems and was a major factor in the Bell System being able to dominate the telecommunications industry for 100 years, until microwave radio technologies began to make conventional land lines obsolete.

While the expiration of the original telephone patents permitted the rapid growth of competition during the 1890s, the acquisition of new patents and a remarkably astute and consistent business strategy resulted in the Bell System's control of more than 80% of the telephone market during most of this century.

During this period the prevailing view of federal regulators, a view influenced by AT&T, was general accommodation to the company's plans in return for promises by the company to provide reliable, economical, universal telephone service. AT&T kept its part of the bargain, providing the best telephone service in the world on terms that anyone could afford.

The ultimate breakup of AT&T in 1984 was not greeted by universal applause, most customers and regulators expressing concern and doubt about the outcome. Years after the divestiture, AT&T still dominates the long distance market by the voluntary choice of its customers, even under the handicap of stringent regulation from which its competitors are exempt.

Technological change and a regulatory policy that had become increasingly illogical on economic grounds had created growing resentment on the part of potential competitors innovative ideas who were not afraid to attack the Bell System in court. As telecommunications technology advanced, the argument that telecommunications is a natural monopoly, never entirely correct, became false. AT&T's comfortable business existence was leading to technological stagnation in telecommunications services and products in spite of the great technological expertise available through Bell Laboratories.

Before World War I, the various owners of patents relating to vacuum tube and radio technology through lack of agreement on cross-licensing effectively stopped progress in wireless communications and radio broadcasting. Only the intervention of the U.S. Navy, with a strategic interest in wireless communications during wartime, forced the parties to develop radio technology. Similarly, it was the wartime development of radar by the government, not the industry itself, in World War II that led to great technological advances in the postwar period. There are many heartbreaking stories of important independent inventions in the area of long distance radio communications and FM radio that were shelved for business reasons, ruining the lives of the inventors.

The federal government is the primary source of funds for university research in science and engineering, although many state governments are becoming active supporters of these fields. In spite of an interest in promoting science and technology on the part of federal and state governments, the legal and regulatory systems favor decisions preserving the economic status quo to the disadvantage of innovative new businesses, and many inventions with economic promise make it to market only after years of delay, if at all. Similarly, U.S. policy on foreign trade can have a dramatic effect on innovations through changes in domestic markets and through changes in export opportunities.

At the beginning of the 1990s, existing and potential future technological advances in telecommunications are so great that the historic tendency for regulators to preserve the status quo is crumbling. The very definition of the telecommunications market is changing as the newspaper and publishing industries find themselves, as heavy users of telecommunications technologies, confronted simultaneously by competing electronic information services.

Television, similarly, faces a severe shortage of conventional broadcasting channels while new telephone technologies provide the potential for large numbers of simultaneous transmissions of programs over broad-band opticoelectronic cable systems, perhaps in tandem with other information services. These developments are viewed with dismay by the cable television industry in the United States, but the reverse is happening in Britain where U.S. Bell companies owning cable TV systems are threatening to deliver telephone service over the same cables.

Other advanced countries are moving rapidly ahead in telecommunications, believing that advanced communications are critical to economic growth and political power in the future, as well as essential for a well-educated, productive work force. As telecommunications services improve, so does the potential for reduction in individual liberties. An informed public must make up its collective mind concerning the trade-offs between the advantages of rapid telecommunications progress in competitive markets sometimes accompanied by unpleasant side effects, or a well-regulated but stagnant telecommunications industry offering no surprises.

CHAPTER 2

A Brief History of Telecommunications in the United States

The quest for rapid communications over long distances was first realized with the development of a visual telegraph system in late eighteenth-century France. Designed by the engineer Claude Chappe, the first link was 230 kilometers long and connected Paris with Lille. A series of towers were located on hills in sight of one another. The letters of the alphabet and numerals were represented by movements of wooden arms mounted on top of a tower, and these movements could be seen by telescope at the next tower and transmitted to a succeeding tower by the same means. Initial success quickly led to a network of such towers covering most of France. This early optical telegraph system proved invaluable to France during the Revolution because it permitted the rapid deployment of military forces in response to allied attacks launched from many quarters (Brown 1970, 12–13). Other countries, noting the military advantage conferred by this optical telegraph system, quickly copied it.

The use of electricity to transmit messages over long distances attracted interest in the eighteenth century, but it was the middle of the nineteenth century before electric telegraph systems began to come into use. Having heard of European research on the use of electricity to communicate over long distances, Samuel Morse became interested in the subject, and he constructed a working model of a telegraph in 1836. Morse applied for a patent on his telegraph device in 1838. At about the same time in Europe, the first electric telegraph systems were being constructed along railway tracks. A working system linked Paddington Station in London with West Drayton, thirteen miles away, by 1839.

After much difficulty, Morse secured a $30,000 government grant to build an experimental line between Baltimore and Washington. The line was operated by the Post Office during 1845, but was not a commercial success. Morse and his backers then tried, unsuccessfully, to sell the rights to the telegraph to the government, and finally began to license the technology to a number of small companies as commercial interest slowly grew. Competing telegraph lines, based on different patents, were soon constructed in many areas in the United States, often along railroad rights-of-way.

After litigation over patent rights, numerous small telegraph companies were gradually forced to merge with the Morse organization, resulting in the formation of the Western Union Company in 1855. Although competitors owning rights to different patents, such as the American Telegraph Company, continued to build and operate lines, Western Union eventually dominated the national telegraph market.

The potential for long distance communication between countries by telegraph was obvious at a very early stage, and the development of submarine cable technology in Europe permitted the construction of a cable connecting Ireland and Newfoundland in 1858, followed by cables crossing most of the oceans before the end of the century.

In 1857, led by the American Telegraph Company and the Western Union Company, the U.S. telegraph companies controlling relevant patents formed a cartel providing exclusive connecting and patent sharing arrangements with members and refusing such arrangements to outsiders (Brock 1981, 73–82). The formal telegraph cartel was consolidated by the acquisition of the last major independent company, the Smith-Kendall line between Washington and Boston, in 1859. In 1860 the Western Union Company, breaking with the cartel over financing arrangements, constructed its own telegraph line to the Pacific Coast. The great profitability of this new line strengthened Western Union and encouraged the breakup of the telegraph cartel during the next few years.

As the original Morse patents were expiring in 1860 and 1861, leaving the cartel members open to competition from new firms, the demand for telegraphic services was increasing dramatically due to the Civil War. The Western Union Company, whose lines were mostly located in the northern states prior to the War, benefited greatly. The manager of Western Union was appointed Chief of U.S. Military Telegraphs for the duration of the war. The 15,000 miles of telegraph lines built by the Union forces were turned over to the Western Union Company after the war, to the great commercial advantage of that company. By merging with American Telegraph Company, its principal competitor, Western

Union established a monopoly in the domestic telecommunications industry in 1866 (Brock 1981, 82–83).

Communication by electric telegraph was not completely satisfactory because only one message could be transmitted at a time, at a rate of one letter at a time, and several hours might elapse before a reply was received. Telegraphic communication could be improved if several messages could be transmitted simultaneously, and methods for accomplishing this began to appear in the 1870s. J. B. Stearns invented a duplex method for transmitting two messages simultaneously in 1868 (Brock 1981, 87), and this method was in regular use by the Western Union Company in 1872. Thomas Edison and George Prescott developed a four-message (quadruplex) telegraph system that was in widespread operation in 1878. During the 1870s, Elisha Gray, an employee of the Western Electric Company, the partially owned manufacturing subsidiary of Western Union, was working on a system for transmitting as many as 15 simultaneous messages using musical tones.

Gray realized that such a system could be used to transmit speech and filed a preliminary patent application, or caveat, on February 17, 1876, for a "harmonic telegraph." Gray had not actually demonstrated speech transmission using such a device. At about the same time, Alexander Graham Bell was also developing a "multiplexing" telegraph for possible use in teaching the deaf to speak (Wasserman 1985, 16–17). Thomas Sanders and Gardiner Hubbard, fathers of two of his students, provided financing for Bell's experiments in return for an interest in any commercial developments. Bell realized that a multiplexing telegraph might be able to transmit speech and, without demonstrating such a device, filed a patent application earlier on the same day that Gray filed his caveat. Bell received a patent entitled "Improvement in Telegraphy" on March 3, 1876, and demonstrated speech transmission on March 10, 1876. Bell ultimately used a device similar to that described in Gray's caveat, and this, plus the close timing of the two patent applications, as well as the confusion over whether the patent applications were for telegraph or telephone systems, led to later litigation between the Bell group and Western Union (Brock 1981, 90). By 1877, Bell and his associates had formed the Bell Telephone Company, which manufactured and leased telephones to customers who were responsible for constructing their own lines between pairs of phones.

A Western Union subsidiary had been providing profitable telegraph services to New York brokers at the time the telephone was being developed. To protect this specialized market Western Union formed the American Speaking Telephone Company as a subsidiary in 1878. At this time both the Bell group and Western Union were rapidly expanding telephone service in order to gain competitive advantage. The introduction of switching centers improved telephone service but required large investments in wire and equipment. Due to a lack of capital, the smaller Bell group resorted to franchising local telephone companies in potentially lucrative markets. The New England Telephone Company, formed by the Bell group in 1878 to promote the construction of telephone exchanges, paid 50% of its stock to Bell Telephone, from whom it also rented telephones. Meanwhile, Western Union with its extensive telegraph system and greater financial resources, was establishing telephone exchanges around the country.

Theodore Vail, appointed general manager of Bell Telephone in 1879, continued the program of expanding telephone exchanges while improving quality of service through technological advances. As another sort of competitive behavior, Bell Telephone, concerned about the rapid expansion of the Western Union telephone system, filed a patent infringement suit against Western Union in 1879.

The smaller Bell Company, starved for capital, found it difficult to compete with Western Union's combined telegraph and telephone system. Western Union refused to provide telegraph service to locations that installed Bell telephones, and Bell's growth was slowing.

In November 1879, Bell and Western Union settled their dispute out of court. Western Union agreed to give up its telephone patent claims and withdraw from the telephone business. Bell agreed to pay Western Union royalties on Bell Telephones over the 17-year life of the agreement and to stay out of the telegraph market. Western Union sold 56,000 telephones in 55 cities to Bell. This agreement was immensely significant to both companies because, at the time, telegraph was the only practical system for interconnecting distant exchanges. It appeared that, in return for giving up its local telephone business, Western Union would have a monopoly on the long distance market. The two companies would act as a duopoly, providing complementary services under an agreement that eliminated competition between them (Brock 1981, 96–99). However, Vail insisted on Bell retaining the right to develop long distance telephone service in the future, should the technology permitting such a service become available.

Thus, Bell Telephone, through its agreement with Western Union and its control of important patents, had established a textbook telephone monopoly that continued to exist until 1894. During the monopoly period, Bell expanded into profitable urban markets but kept prices high and ignored the smaller cities and towns, and so total telephones in use grew more slowly than they did in the previous competitive period. There were about 142,000 phones in service in 1885 and 270,000 phones by 1894. The unsatisfied demand for local telephone service in smaller communities and rural areas led to the formation of thousands of local telephone companies in the late 1890s as the Bell patents expired.

In 1880, Bell had been reorganized as the American Bell Telephone Company. Anticipating the expiration of the original patents in 1893 and 1894, Bell sought to erect barriers to entry through an active engineering program that obtained hundreds of new patents on all aspects of telephone equipment, through the establishment of long distance telephone service which the new local companies could not provide, and through the vertical integration of manufacturing and telephone operations.

Technical developments were beginning to improve long distance telephone service, and a subsidiary, American Telephone and Telegraph Company (AT&T), was formed in 1885 to serve the new market. In order to implement its vertical-integration strategy, Bell acquired an interest in the Western Electric Company, the manufacturing subsidiary of Western Union. Bell's interest was later increased to 100% ownership of Western Electric. Bell actively followed a consistent strategy of monopolizing all aspects of the telephone business in order to deny possible new entrants a foothold in any part of the business. In 1899, prevented from expanding by Massachusetts law, American Bell Telephone Company transferred control to its New York subsidiary, AT&T, which was to control the Bell system until 1984 (Brock 1981, 118).

By the late 1890s, the original patents having expired, hundreds of new local telephone companies and cooperatives were being formed each year in the United States. Spurred on by attractive profits and not deterred by Bell's hundreds of recent patents, more than 4,000 local systems had come into existence by 1902, mostly located in small towns in the midwest.

Bell's share of the telephone market had fallen to about 56% of the telephones in the country in 1902 and to 51% by 1907. Bell brought patent infringement suits against some of its new competitors, but Bell's recent patents were either narrowly interpreted or invalidated by the courts in many cases, and the suits did not discourage competition. Bell continued to fight the new competition by reducing prices, by improving and extending its long distance system, and by selective mergers with competitors. However, while the Bell System lost market share during this period of competition, the market for telephone services was growing rapidly. While the number of telephones in use had grown at a rate of

6.3% during the decade of the Bell monopoly, the number of telephones grew at a rate of 21.5% during the decade of competition.

In the 1890s AT&T engineers, familiar with the research that British scientists such as James Clerk Maxwell and Oliver Heaviside had performed on inductance in electrical transmission systems, were actively searching for methods for improving the efficiency of long distance telephone service. Although many others were engaged in closely related work, Michael Pupin, Professor of Physics at Columbia University, patented the loading coil in 1900 and sold the rights to the Bell system shortly after (Wasserman 1985, 13). This simple invention, by increasing the inductance of transmission lines, increased the range and quality of long distance telephone communication while lowering system cost by permitting thinner wire to be used.

Although enjoying the advantages of system size, quality of service and long distance connections, the American Bell Company was seriously threatened by competition in the early 1900s. In spite of rapid expansion, refusing connections to independents, price reductions and threats of patent litigation, the Bell Company was encountering difficulty keeping new entrants out of the long distance market. Profits were declining, and managerial inefficiencies were becoming apparent.

The European governments had organized national telecommunications companies, and there existed the very real possibility of a similar policy of nationalization in the United States. In the face of these various competitive and governmental threats, Theodore Vail established a policy of submitting to government regulation while retaining private ownership and control of the Bell System, and this continued to be company policy throughout its history, eventually serving to keep competitors out of the industry as the technology changed and the rationale for a monopoly in telecommunications disappeared.

The merger policy and an increasing demand for long distance connections reversed Bell's decline in market share. The independents' share of the market decreased from 49% to 42% during the 1907–1912 period. The growth of the telephone market slowed as the industry matured and prices stabilized, with mergers rather than new telephone installations increasing Bell's share of the market. Strategic mergers with smaller firms weakened other interconnected independents and allowed them to be purchased separately on advantageous terms.

But the merger policy was not risk free. Possible Sherman Act violations forced Bell to negotiate a settlement with the attorney general in 1913. In the Kingsbury Agreement, named for a Bell vice-president, Bell disposed of its Western Union stock, allowed interconnection with independents, and agreed to refrain from acquiring more independent companies. Thereafter, the number of Bell system telephones increased substantially, and Bell's market share had grown to 79% by 1932.

The patenting of radio communication by Guglielmo Marconi in 1896 and the formation of the Marconi Wireless Telegraph Company in the United States in 1899 constituted a new technological threat to the Bell System (Brown 1970, 22). The Marconi Company monopolized wireless telegraph communications in the United States and Britain, posing a competitive threat to conventional telegraph, the telephone and undersea cables. The Marconi invention was widely used for communications at sea by 1907, and the U.S. Navy immediately recognized its military significance.

Bell responded by expanding its research organization and its emphasis on basic science. Western Electric also increased expenditures on research, partly for war reasons, its R&D department increasing from 192 engineers in 1910 to 959 in 1915. During this period AT&T carried research on new inventions up to the point of patent applications and then discontinued work. The acquisition of patents on technologies that Bell did not intend to use served the dual purposes of hampering technologically based competition and providing something to trade to others who possessed patents desired by Bell as improvements on existing telephone technology.

Since other companies were following similar strategies, the rate of introduction of telecommunications innovations slowed. As an example, the basic patent on radio communication was held by the Marconi Company, but Lee DeForest had patented the three-element vacuum tube which, as an efficient amplifier, was necessary for improvements in radio communication. DeForest's company introduced wireless communications to the Navy, but later failed after being sued by the Marconi Company over patent infringement. DeForest sold his patent rights to AT&T in 1913. The Marconi Company had purchased rights to the Fleming patent on a two-element tube used for radio signal detection. Both types of tubes were necessary for efficient radio communications, but the court ruled in 1916 that each device was an infringement on the patent of the other when used for radio, thus preventing either company from commercializing the two inventions. AT&T attempted to improve on the DeForest tube by designing a high-vacuum version suitable for use as a telephone amplifier, but the General Electric Company had patented high-vacuum tubes. General Electric also possessed the rights to still another invention necessary for efficient radio communications, the Anderson alternator, which was used to generate radio waves. Finally,

Westinghouse controlled the patents to the Armstrong feedback circuit, also an important component of radio technology. The complex patent situation, combined with incentives for companies holding the patents to protect their existing markets, delayed the commercialization of improvements in radio technology. Armstrong later committed suicide after years of legal wrangling that intentionally delayed the introduction of his FM technology.

With the coming of World War I and the cutting of the transatlantic cable between Germany and Britain in 1914, radio became militarily important. At the time, radio communication was actually wireless telegraphy using Morse Code. The Marconi Company had a monopoly on radio devices, manufacturing 95% of such equipment in 1917, and the patent situation was hampering the development of radio. The U.S. Navy took responsibility for possible patent infringements during the war and ordered all interested companies with the required expertise to produce radio equipment.

The Navy constructed and operated ship-to-shore voice communication systems and transatlantic radio networks during the War, but it was required to terminate its coastal radio communication activities after the armistice. After World war I, the companies in the radio equipment industry returned to the prewar patent impasse. With the support of the Navy, still a large customer for radio equipment and still interested in the military aspects of radio technology, General Electric (GE) purchased the assets of the American Marconi Company and formed the Radio Corporation of America as a subsidiary in 1919.

In 1920, GE and AT&T agreed to cross-license vacuum-tube technology. The agreement licensed uses of the technologies rather than the rights to the patents, thus delineating the markets served by the two companies (Brock 1981, 166–67). AT&T received exclusive wire telephone and wire telegraph rights while GE received exclusive wireless or radio telephone rights in specialized markets, radio broadcasting being still in the future. A similar licensing agreement was reached with Westinghouse the next year, accompanied by stock purchases and equipment purchase arrangements. All parties benefited by the technology licensing agreements, which provided each company a satisfactory level of market protection from the others and formed a legal wall preventing intrusions by outsiders. In compensation for trading its rights to use the aspects of radio technology it had developed in its own business, AT&T received assurance that its near monopoly of long distance wire telephone service would continue. The agreements thus delayed for decades the development of long distance wireless telephone communications, a much more efficient technology.

But a new business, not contemplated by the 1920 agreements, began to emerge as the result of a Westinghouse employee playing records on a phonograph connected to a transmitter being developed for the Navy. Amateur radio enthusiasts, listening to the transmissions of music on their homemade receivers, generated an interest that resulted in Westinghouse initiating a radio station in Pittsburgh in 1920. The station was intended to stimulate interest in Westinghouse radio equipment, and was so successful that radio stations were soon in operation in many cities. RCA sales of radio receivers exceeded $11 million in 1922, attracting the attention of AT&T.

AT&T entered the new market by establishing radio stations connected by wire telephone lines, believing that radio broadcasting was a natural adjunct to its wire telephone monopoly. RCA called for arbitration, believing that both AT&T's sale of radio receivers and its refusal to connect RCA stations to be violations of the 1920 agreement.

In 1926, after the arbitrator had ruled in favor of RCA, AT&T agreed to abandon radio broadcasting and receiver sales in return for a stronger market position in two-way communications. RCA purchased AT&T's broadcasting assets, which became a part of the National Broadcasting System, and agreed to use AT&T's wire system to interconnect radio stations. It was now in AT&T's interest to provide connecting services to RCA because it was no longer in the broadcasting business and wished to maintain its monopoly of wire telecommunications. The 1926 agreement also divided rights to the use of the future television technology between the two companies.

The 1926 agreement was seen by the government as a conspiracy in restraint of trade in violation of the Sherman Act, and antitrust charges were filed against GE, Westinghouse, and AT&T in 1930. The suit was settled in 1932 by the defendant companies consenting to eliminate the exclusive provisions of the cross-licensing agreement. GE and Westinghouse were required to sell their RCA stock. The consent decree did not affect the operations of the companies in any material way, merely limiting the parties to their existing markets. The commanding technological and market positions of the existing companies were believed by the parties sufficient to guarantee their continuing dominance of those markets.

AT&T's research on motion picture equipment during the 1930s provided additional technological rights to be traded away in protection of its telephone business. AT&T believed, during this period, that it was necessary to pursue R&D activities well into any potential competitor's field in order to be in the most advantageous position to deter such potential competitors from entering the telephone market. By the middle 1930s AT&T owned about 9,000 patents and was licensed under about 6,000 additional patents owned by other companies. The cost of the R&D program that built this legal wall around AT&T could be charged as an expense to the operating companies and recovered from telephone customers through rate regulation procedures. By 1934, AT&T owned about 80% of the telephones in the United States, and only the threat of antitrust charges under the Sherman Act had kept it from playing an even larger role in this market and perhaps occupying similar positions in other communications markets.

Between 1934 and 1956 very little changed in the structure of the telecommunications industry. AT&T remained the largest company with about 80% of the telephone market and control of the long distance system, and barriers to entry were still formidable. The strategic value of the national telephone system during this period of alternating international tension and war was recognized by the Department of Defense, which became a supporter of the Bell System. However, the legal wall surrounding Bell's telephone business gradually changed during this period, losing its technological foundation as patents expired and acquiring a regulatory foundation.

Theodore Vail's original policy of agreeing to government regulation as long as it was fair matured into close working relationships with the Federal Communications Commission and the various state regulatory commissions. These relationships formed a protective wall around AT&T's telephone business every bit as sturdy as the former patent barriers, replacing market forces with regulatory decisions made within the political context of the 1930s.

The Depression caused people to seek a greater role for government in the economy. In 1934, President Roosevelt, among many other initiatives, asked Congress to consolidate the roles of several existing regulatory agencies in one new agency, the proposed Federal Communications Commission. The FCC, authorized by the Communications Act of 1934, was granted new tariff authority over long distance service as well as the authority to require interconnections, approve expansions and allocate frequencies. A "Special Telephone Investigation" was ordered by Congress, and while it was being conducted, the FCC, under the presumption that existing long distance telephone rates were reasonable, instituted a process of continuing telephone rate surveillance rather than requiring the company to periodically justify its rates (Henck and

Strassburg 1988, 4–5). Informal negotiations between the FCC staff and representatives of AT&T replaced official rate hearings.

AT&T voluntarily reduced its rates from time to time, as its costs declined through technological improvements or falling prices in that deflationary era, and the FCC took credit for the rate reductions even though they might well have been made voluntarily by the company in response to market forces. The first official telephone rate investigation by the FCC did not occur until the 1960s.

In 1938, upon conclusion of the "Special Telephone Investigation," a proposed report of the FCC recommended that AT&T be divested of its Western Electric subsidiary and that the operating telephone companies be required to use competitive bidding when purchasing supplies and equipment. The operating companies were subject to rate regulation, but Western Electric was not. Its unregulated equipment prices could be passed on to the operating companies and thence to their customers as legitimate costs under rate regulation. AT&T lobbied extensively to retain the Western Electric tie, and the final FCC report in 1939 did not recommend dissolution.

The development of television and microwave technology constituted possible competitive threats to AT&T in the postwar years. AT&T had been providing connection services to radio stations since the 1920s, but the conventional copper-wire telephone line was not capable of carrying the much more information-intensive television signals. The transmission of a television program required about the same capacity as a thousand simultaneous telephone conversations. Television programs could be transmitted over coaxial cables, developed in the 1930s to carry multiple telephone conversations, or over microwave radio systems. A physical path for transmission of the signal is not necessary in a microwave relay system, only towers at intervals of 20 to 40 miles. Therefore the cost of long distance transmission of information, especially high-volume business communications or high-density information such as television programs, is significantly reduced using microwave systems.

Interest in satellite communication systems was also developing in the 1950s, and the first active communications satellite was put into operation in 1960. The FCC concluded in 1961 that AT&T was best qualified to operate a global satellite communications system, but there was significant opposition from a vocal minority opposed to expanding AT&T's monopoly out into space at government expense. The Communications Satellite Act of 1962 established the Comsat Corporation as a competitor and vendor to AT&T.

The original radio patents had either expired by the end of the war or lawsuits such as that of Zenith against RCA, settled in 1957 in favor of Zenith, broke down the patent barriers to entry. Because microwave radio is an application of radar technology and many companies had worked on radar during the war, potential new entrants did not face patent barriers preventing use of the new technology. However, microwave frequencies were allocated by the FCC, and the allocation procedures could be used, and were used, by established telecommunications firms to deny entry to new firms.

The FCC began by freely giving temporary experimental licenses to companies wishing to use microwave technology, but the uncertainty over what would be the FCC's final policy on permanent licenses and rights deterred the large investments required by the new firms. Philco, Western Union and IBM/GE were early entrants, establishing microwave links between New York, Philadelphia and Washington in the 1940s. Raytheon was licensed to operate a Boston–New York–Chicago microwave system.

AT&T had previouly planned the exclusive use of coaxial cable systems for high-volume long distance communications. However, aware of the advantages microwave technology could confer on competitors, AT&T began operating a New York–Boston microwave system in 1947 and gave the new technology high development priority. AT&T requested that the FCC allocate microwave frequencies exclusively to common carriers on the basis that the carriers were not permitted to engage in other businesses. This request was at first denied, but by 1948 the commission had decided to allocate microwave frequencies only to common carriers. This decision, and an interpretation of temporary licenses as intended for experimental use only, effectively excluded new competition from the microwave telecommunications market. The most viable potential competitors were heavily involved in other businesses and could not meet the definition of common carrier. AT&T filed a video transmission tariff in 1948, providing services eight hours per day, seven days per week. The tariff prohibited interconnection with non-AT&T transmission systems, thus excluding the other important common carrier, Western Union, and providing AT&T with a monopoly over the video transmission business.

The AT&T connection prohibition was attacked by the television broadcaster's association in FCC hearings in 1948. The FCC decided in 1949 that AT&T must permit connections with broadcast company communication systems, but did not have to provide connections to Western Union, a competing common carrier.

While the FCC left open the option of a common carrier establishing a national microwave system for video relay separate from that of AT&T, the enormous investment required and the uncertainty over permanent license allocations effectively deterred such competition. The FCC believed that it was in the public interest to permit only one national microwave system in the important video transmission market because of the assumed shortage of microwave frequencies. The FCC accepted the argument that AT&T was a natural monopoly and therefore could serve the market at lower cost than could a group of competing firms with duplicate facilities. The result was relatively slow development of microwave communications in a period of growing demand for such services. Since AT&T was guaranteed all such business, there was no urgency in expanding to meet the demand. In the 1950s, as the size of the market became clear, it was realized that frequency availability was better than previously supposed.

The obvious profitability of the new communications technologies combined with a strong and growing demand for microwave services attracted new firms to the long distance telecommunications business. The FCC began a study of the allocation of the microwave spectrum above 890 MHz in 1956, and, in spite of the active opposition of the common carriers, established a liberal frequency allocation policy in its final report of 1959. "Theoretically, just about anyone was now eligible to build a private microwave system" (Henck and Strassburg 1988, 84).

Private microwave networks were constructed and operated by businesses with a high volume of internal long distance communications, thus bypassing the Bell System. An extremely important factor in the entry of competitors and bypass systems into the high-volume long distance market was AT&T's regulated price structure based on the average cost principle, which did not enable the new technology-based economies of the common carriers to be passed along to customers in the form of lower rates. Thus, potential competitors saw great economic opportunities in the microwave communications business. The stage was set for the entry of competition into the interstate telecommunications market.

CHAPTER 3 _____

The Origins of the AT&T Divestiture

In the 1950s important telecommunications regulatory issues began to take form. The telephone terminal equipment market was structured in a manner similar to that of the telephone service market. AT&T, the dominant force in the service market, purchased essentially all of its terminal equipment from its subsidiary, the Western Electric Company. Telephones owned by the Bell System were installed in businesses and homes, and all maintenance was performed by Bell System employees. General Telephone, the next largest U.S. telephone company, had a similar manufacturing arrangement with its subsidiary, the Automatic Electric Company. The market for terminal equipment for the remaining independent telephone companies was too small to encourage independent terminal equipment manufacturers to enter the business. AT&T argued that the mandate for a safe, reliable, high-quality national telephone service required that no "foreign equipment" connections to the telephone network be permitted. Federal and state regulators, generally accepting the argument that the telecommunications system was a natural monopoly, tended to support AT&T's determination to provide all customer telephone equipment without exception.

Occasionally a firm would market a product to be used in conjunction with conventional telephones, and when this happened, AT&T responded with threatened or actual legal action. Since 1921, the Hush-A-Phone Company had been marketing a simple cuplike device to be attached to a telephone set, thus providing privacy to the speaker. Over the years AT&T had informed users of the Hush-A-Phone that the device violated tariff restrictions, and the Hush-A-Phone Company finally filed a complaint against AT&T with the FCC in 1948. After seven years of hearings and deliberation, the FCC decided against the Hush-A-Phone Company in 1955. The FCC's decision was supposedly to safeguard the quality of telephone communications, but in fact it was recognized that the existing policy of one national integrated telecommunications utility was in question, and no foothold would be permitted to competitors, no matter how benign their intentions.

Hush-A-Phone appealed the FCC decision in court and won. The court accepted the argument that the product provided benefits to the user and was not detrimental to the telephone system. The principle that some public harm must be shown in order to justify restrictions on customer-owned accessories to Bell telephone equipment became accepted by the FCC. However, AT&T took a very narrow interpretation of the court's decision and continued to prohibit the connection of "foreign" equipment other than simple devices such as the Hush-A-Phone.

In the late 1950s, in response to a perceived demand by oil industry users of mobile radios in the southwest, the Carter Electronics Corporation was formed to market the Carterphone, an acoustic device for connecting a mobile radio to a telephone handset. There was no electrical connection between the telephone handset and the Carterphone, but the device violated the AT&T tariff because it connected the telephone system to another channel of communication. AT&T threatened suspension of service to users of the Carterphone. The FCC, in accordance with the principle that the Carterphone did no harm to the telephone system and benefited users, declined to forbid its use.

In the face of continuing AT&T resistance, Carter filed an antitrust suit against AT&T in a Texas court. The court, unsure of jurisdiction and other issues, referred the matter back to the FCC, which began an investigation in 1966. In 1968 the commission found that the Carterphone was not detrimental to the telephone system, and that AT&T's tariff was illegal because it violated the court's earlier ruling on telephone attachments.

AT&T filed a new tariff requiring the use of AT&T-supplied protective connection attachments (PCAs) when connecting "foreign" devices to the telephone system, but this tariff drew many complaints that it also violated the court's ruling. The connection devices were available from AT&T on lease terms that eliminated any economic incentive to market innovative terminal telephone products. A PCA was required when a customer connected a purchased device to the telephone system, but not when the identical device was obtained through the Bell system.

In the 1970s a virtual technological explosion provided an enormous variety of terminal equipment for use on the customer's premises. The key regulatory issue was the extent, if any, the common carrier providing telephone services was to be permitted to monopolize communications facilities on private premises, keeping customers from benefiting from innovative products.

While the FCC had long supported the concept that AT&T's monopoly was in the public interest, the Department of Justice was not convinced, filing an antitrust suit against AT&T and Western Electric in 1949. This suit was based on information originally developed during the FCC "Special Telephone Investigation" of the 1930s. The government sought to separate Western Electric from AT&T and to end all restrictive agreements between AT&T, Western Electric and the Bell operating companies. The relationship among the Bell companies was described as a price-fixing conspiracy establishing Western Electric as an unregulated monopolist in the telephone equipment market.

AT&T enlisted the support of the Department of Defense in its argument for the retention of Western Electric, the Korean War having begun while the litigation was in progress. AT&T denied the antitrust charges, asserting that its relationship with Western Electric was necessary for efficient operations, and argued for a settlement agreeing to restrictions in return for protection from future antitrust charges. The suit progressed slowly, finally resulting in a consent decree in January 1956.

AT&T succeeded in maintaining its integrated organization, but at the price of more-restrictive government regulation and a quasi-public business status. AT&T's argument that its regulated status exempted it from the antitrust laws was accepted, but the Bell system was prohibited from expanding beyond its established telecommunications businesses, and it was forced to license its patents and provide technical information, at reasonable cost, on request.

AT&T had actually been following a liberal patent and technical information policy prior to the settlement, one of the indications that AT&T considered itself to be a model corporate citizen. Many of AT&T's most important patents were made available royalty-free since, as a regulated monopoly, patent barriers to entry were not required. Once again, AT&T had traded away technological rights for assurance that its own markets would be safe. Among the technological rights traded away were the right to control transistor applications through licensing agreements. The consent decree encouraged many companies to begin manufacturing transistors with a consequent drop in transistor prices. The widespread availability of inexpensive transistors lead to the worldwide solid state revolution in the electronics industry, a revolution still in progress.

By 1956 AT&T had in place regulatory barriers to entry that guaranteed its markets, especially the more and more lucrative long distance market. Regulatory delay made challenges to AT&T's position extremely costly. Such challenges, even if successful, were risky since AT&T would have time to prepare a business response to prospective competitive activities before they were approved by the regulators. Finally, potential competitors in the long distance market had no way to connect their competing service to local subscribers' equipment.

Microwave technology is most suitable for high-volume long distance communications, the cost per circuit dropping rapidly until circuits in use exceed 250 and then continuing to fall slowly until about 1,000 circuits are in use. At low levels the costs are due mostly to property acquisition and transmission facilities and at high levels mostly to multiplex equipment. Therefore, there are extensive economies of scale in microwave communications up to about 1,000 voice circuits, and the economics of microwave technology limits new entrants to high-density routes, forming an entry barrier of sorts.

In spite of these barriers to entry into the microwave communications industry, AT&T's pricing scheme attracted competitive interest because it was not based on cost. While AT&T's rates for long distance service had been slowly falling in response to technological advances over the years, they were still based on conventional technology during a period in which the expanding microwave systems were rapidly reducing costs, especially on high-volume routes.

At that time AT&T did not give volume discounts, its long distance rates being based on the average cost of all of its installed equipment rather than on marginal cost, the latter defined as the much lower cost of the microwave systems actually being employed on high-density routes. AT&T's pricing philosophy was called *national rate averaging*, and was generally applauded by regulators since such long distance rates were perceived as being fair to all classes of customers. Rate averaging subsidized high-cost rural service with revenues from low-cost, high-density areas, and this policy was politically popular.

AT&T depreciated its installed equipment slowly, the higher asset costs carried on the books being translated into higher rates under the prevailing form of regulation which permitted a fair rate of return on investment. Productive equipment normally falls in value and is replaced when it is no longer competitive with modern equipment. Consequently, while AT&T was only marginally reducing the rates charged for long distance service over modern microwave radio networks, a widening gap began to appear between the true economic cost of providing such service and the rates charged, which were based on the regulator's definitions of allowed costs.

AT&T, in response to increasing pressure from state regulatory commissions and not at all in conflict with its strategy of monopolizing all aspects of the domestic telecommunications market, had agreed to allocate a growing portion of the costs of local telephone plant and equipment to the long distance service.

The proposed methods for computing the allocations were controversial, but such allocations were increasingly based on the relative use of local telephone facilities for long distance and local calls. State regulatory commissions were grateful for the resulting politically popular lower local telephone rates that were being cross-subsidized through separations formulas by funds earned in the profitable long distance market. Such cost allocations to the long distance market provided further justification to the regulators for the slow decrease in long distance rates, in spite of the growing divergence between such rates and the true cost of providing such service. Similarly, a growing divergence, or subsidy, began to appear between the rising cost of providing local telephone service, which was not affected by technological advances at that time, and lagging increases in local rates.

The directions of divergence between costs and rates in the two markets were justified by the Ramsey (1927) pricing theory, which supports the concept of setting a higher price in the less elastic market, long distance in this case, and a lower price in the more elastic market, here assumed to be local telephone service, all in the interest of ensuring that a natural monopolist obtains sufficient revenues to cover all costs.

The resulting loss of earnings in long distance service simply appeared as gains in the earnings of the operating companies, the overall effects on AT&T essentially amounting to internal accounting transactions. The AT&T system gained a stronger position in both markets, assuming continuing legal barriers against entry by competitors, while accommodating the desires of regulators. Among the unfortunate consequences of this pricing philosophy was the elimination of incentives for innovation in local telecommunications markets, because no unsubsidized competitor employing some new technology could match the Bell System's local rates.

In 1956, as technological progress made it clear that the availability of microwave frequencies was better than previously believed, the FCC reconsidered the allocation of microwave frequencies to companies that were not common carriers. The Bell System's microwave transmission capability then acounted for 22% of the 10.5 million miles of telephone circuit capacity and 78% of the 60,000 miles of intercity television capacity. Western Union employed microwave transmission in its telegraph service between New York, Washington and Pittsburgh. Private microwave systems operated by utilities and petroleum companies along rights of way totaled 31,000 route miles.

Several hundred individuals and organizations attended the FCC hearings at which AT&T, Western Union and the independent telephone companies argued that all microwave frequencies should be allocated to common carriers in view of their extensive existing investments and status as regulated utilities serving the public interest.

As mentioned earlier, the Bell System pricing philosophy was based on the idea of nationwide rate averaging. The same rate that generated profits on a high-density, low-cost route provided a subsidy to customers in a low-density, high-cost area. AT&T argued that competition would weaken its ability to provide a high-quality nationwide communication system because competitors would choose to enter only the most profitable markets, engaging in "cream-skimming" instead of offering across-the-board competition and depriving the Bell System of the revenues necessary to support continuing service to unprofitable areas. The Bell witnesses, using the natural monopoly argument, claimed that the duplication of communication facilities would also increase customer costs, since essentially the same pool of customers would pay rates that must cover all the costs of duplicate systems.

Microwave equipment manufacturers and potential private-system operators argued that adequate frequencies were available and that there was no reason for the FCC to protect the economic interests of the existing common carriers. They sought permission for the sharing of microwave system capacity by groups of private users, an economic necessity due to the extensive economies of scale in microwave communications. Technical data contained in a very extensive study prepared by the Electronic Industries Association indicated that hundreds of microwave stations could be accommodated without interference in a given metropolitan area. Since only Los Angeles had as many as 38 microwave stations, it was argued that there was plenty of unused microwave frequency capacity in U.S. cities.

In July 1959, the FCC concluded that adequate microwave frequencies existed and made microwave frequencies generally available to private users upon application. The FCC did not authorize the sharing of private systems, a detriment to the implementation of such systems due to the economies of scale referred to previously. Requests for permission to interconnect private systems and common carriers were to be considered on a case-by-case basis. Only a private user with a very high volume of long distance communications would find the introduction of a microwave system advantageous. The FCC decision did no direct harm to AT&T, but it opened the door to possible future harmful actions by competitors; AT&T petitioned unsuccessfully for a reconsideration of the FCC's decision.

Since AT&T's rates for long distance services tended to be uniform with little or no provision for volume discounts, AT&T could deter localized entry into the long distance microwave market by reducing the rates charged to those companies likely to build private systems while maintaining existing rates to all others, an application of price discrimination and a relatively low-cost way to protect existing markets. AT&T promptly filed a new tariff called Telpak, which eliminated incentives for companies to build private microwave systems where AT&T facilities existed.

Telpak provided large discounts to users of groups of private lines. Telpak A offered a discount of 51% for the use of at least 12 lines; Telpak B offered a discount of 64% to users of at least 24 lines; Telpak C offered a discount of 77% to users of at least 60 lines; and Telpak D offered a discount of 85% to users of at least 240 lines. These rates were cheaper than the costs of building private systems with the same capacity, although the rate for a single line was not changed by AT&T. These rates also indicate the astonishing profits that had been made by AT&T using rate averaging on high-volume routes, and this information was not lost on potential competitors. High-volume users gladly accepted the Telpak rates and abandoned plans to build private systems, becoming AT&T supporters in future regulatory hearings.

Western Union, its private-line business being gravely damaged by the Telpak tariff, charged that the tariff was a predatory pricing scheme. Western Union was joined by Motorola, a microwave equipment manufacturer, whose business was also hurt by the tariff because AT&T microwave equipment was entirely manufactured by Western Electric. The FCC conducted an investigation, finding in 1964 that, since the same facilities were also used for single private lines, there was no cost savings to AT&T from selling private lines in bulk through the Telpak tariff.

Thus the Telpak tariff was found to be a case of price discrimination without a cost justification. Telpak tariffs A and B were canceled, and further hearings were held on tariffs C and D. AT&T was pleased at the loss of tariffs A and B because the customers affected, not being large users, were not likely to build their own systems, and AT&T could enjoy the former profits once again. The accumulation of regulatory cost definitions and practices, most if not all of which defy economic logic, hampered the hearings, which continued for years. Ironically, the long-standing preference of regulated utilities for rates based on average cost, having been accepted by the FCC, became the argument used against AT&T in the case of the Telpak tariffs. In 1976 Telpak tariffs C and D were declared illegal, and they were withdrawn in 1977. However, by 1977, many large users, including the U.S. government, supported the Telpak rates, and successfully petitioned for their continuation.

The Telpak volume rates of the 1960s indicated that AT&T's single-line rate was probably far above cost. Could a new firm sign up for one of the volume Telpak rates and resell its private-line rights, one at a time, to individual customers at rates less than the single-line rate charged by AT&T? Could AT&T's Telpak price discrimination scheme thus be defeated by a third party entering and acting as a reseller? Such resale was prohibited by the AT&T tariff, but the obvious economic feasibility of such a plan indicated that a new company, with regulatory approval and interconnection with the AT&T system, could build its own microwave system, thereby providing its own low-cost, high-volume network, and accomplish the same objective.

In 1963, Microwave Communications Inc. (MCI) filed a request with the FCC for authorization to act as a communications common carrier between St. Louis and Chicago. MCI was a tiny company intending to provide low-cost, no-frills private-line communications to specialized users in a restricted area. MCI believed that there existed a submarket for economical long distance services that was being ignored by AT&T. The application was opposed by AT&T and the other common carriers who argued duplication of existing services and lack of interconnection facilities. The FCC viewed the MCI application as an opportunity for an experiment that would provide information on the extent of unserved markets and the viability of alternative communications systems. In October 1967, MCI's application was granted and, after complaints leading to further review, affirmed in 1969. The decision was recognized as possibly having great significance for the future telecommunications industry in the United States, but the majority argument in favor of granting the application was based on only the facts of the case.

The existing regulatory philosophy embraced the idea of a single national regulated telecommunications utility, but the MCI case indicated that changes in that philosophy were imminent. In particular, this was one of the first indications of the growing dissatisfaction with government regulation that was to culminate in the wave of deregulation that occurred during the 1980s. There was a recognition that government regulation might provide a company with an excessively secure, comfortable existence. The regulated company could become a safe haven for inefficiency and technological obsolescence, not required to respond to the changing demands of the public or the actions of potentially more efficient competitors. Under the camouflage of fairness, equal rates were resulting in enormous profits on high-volume routes.

MCI's microwave system, after further legal challenges from the common carriers, opened in 1972. MCI claimed that, while its legal costs over seven years had been $10 million, the cost of constructing its microwave system was less than $2 million, and construction had taken only seven months. These figures give some indication of the barriers to entry into the telecommunications industry at that time, the barriers largely consisting of the lost time and dollar costs of legal challenges. Of course, MCI did not know it would eventually be successful, the uncertainty forming still another barrier to entry.

Following the MCI decision, hundreds of additional applications for specialized microwave systems were quickly filed with the FCC. By April 1970 there were 1,460 applications on file, the number growing to more than 1,700 by July. The number of applications for specialized common carrier status, each application supported by detailed technical and financial information and most applications being opposed by the established common carriers, led the FCC to review the approval process.

The 1968 report of a blue-ribbon panel on communications appointed by President Johnson concluded that regulation of the existing integrated public telephone system should be continued but that all other communications services should be open to competition. Competition was not seen as an end in itself by the FCC. Only if both feasible and likely to result in benefits to the public was competition to be encouraged.

The FCC believed that relatively free entry into the new specialized markets would result in new price and service options, the development of new products and markets, and a faster response to customer needs. The commission streamlined the approval process, granting approval to all specialized common-carrier applicants meeting a modest set of financial and technical requirements. It was not considered necessary to prohibit the new specialized carriers from competing with the established message and wide-area telephone services since the new carriers were not licensed to enter this market.

By 1973, MCI, through expansion and merger with other new companies, was an $80 million dollar system growing rapidly under the management of its new head, William McGowan. The persistently stubborn attacks by tiny MCI on its far larger foe cannot be explained by economic logic. The managers of MCI were simply extraordinarily tenacious businessmen. MCI continued to sue AT&T over allegedly lost business for years, and was usually successful in court. Its opponents called MCI "a law firm with an antenna on its roof."

Existing microwave systems transmitted conventional analog signals, which were satisfactory for voice communications but were inaccurate and inefficient when used for the transmission of computer data. In 1971, Data Transmission Company (Datran), a subsidiary of University Computing Corporation, announced plans for 255 stations, opening the first digital microwave network in 1973 (Henck and Strassburg 1988, 149). AT&T responded by building its own digital network and filing a tariff with attractive rates. In 1976, Datran filed for bankruptcy, filing an antitrust suit against AT&T. In response to complaints to the FCC, the administrative law judge ruled that AT&T's rates were discriminatory, but the rates continued in effect pending the development of adequate cost information justifying the filing of a new tariff. Datran's assets were purchased by Southern Pacific Communications, now a subsidiary of US Sprint. AT&T has continued to expand its digital data network.

In 1970 the Nixon administration announced in a memorandum that the government "should encourage and facilitate the development of commercial domestic satellite communications systems to the extent that private enterprise finds them economically and operationally feasible" (Henck and Strassburg 1988, 153). It is important to understand that while microwave radio relay dramatically reduced the cost of long distance telecommunications because physical paths for bundles of wires strung over large geographical areas were no longer necessary, satellite facilities virtually eliminated distance as a cost factor in such communications.

In June 1972 the FCC announced a domestic satellite policy that, like the policy for specialized microwave systems, provided for relatively free entry. AT&T proposed a joint effort with Comsat in space, but there was concern that AT&T could cross-subsidize such an effort with revenues from regulated businesses or divert existing business to its satellite facility. Both types of action would put competitors at a severe disadvantage and negate the FCC's policy of encouraging innovative approaches to satellite communications. The FCC ruled that for the first three years AT&T could use satellite facilities only to provide regulated public switched telephone services, and that AT&T must divest itself of its 29% interest in Comsat (Henck and Strassburg 1988, 155).

AT&T's response to competitive developments was to seek to isolate the emerging new microwave communication systems by filing tariffs with the state regulatory commissions that included discriminatory rules for interconnection with local telephone exchanges. The complex types of private-line service preferred by the customers of the new competitors were excluded from these tariffs, which permitted only the standard private-line service provided by the Bell System. MCI challenged the AT&T tariffs before the FCC. The FCC had affirmed the principle of no discrimination between competing long distance systems in interconnection arrangements with local distribution services, and decided in favor of MCI in 1973.

AT&T appealed, arguing that the FCC did not have jurisdiction over local interconnection matters since these were regulated by the various state commissions. In 1974 the FCC ruled against AT&T, requiring that the full complement of local distribution services provided to AT&T's Long Lines Department must be provided to MCI on the same terms. By 1976, after some further attempts at restriction of interconnection services by AT&T, the FCC had granted complete interconnection rights to the specialized common carriers. However, five years had elapsed during which AT&T's competitors were operating at a disadvantage, and the introduction of innovative new telecommunications services was delayed.

As the interconnection barriers to entry were being dismantled, AT&T gradually reduced incentives to entry into its most vulnerable markets by rate adjustments. The Series 11000 private-line tariff, reluctantly approved by the FCC on an experimental basis, provided low rates to the same customers that MCI proposed to serve, thus undercutting the new company and signaling to others that AT&T was prepared to compete with price as well as with legal and regulatory actions.

As MCI began service between Chicago and St. Louis in 1972, Western Union quickly filed a tariff matching MCI's prices for specialized private-line services. MCI objected, claiming that the proposed tariff was predatory and might become a precedent for AT&T and other carriers, but the tariff went into effect. AT&T considered a direct competitive price response on this particular route, but filing such a tariff would require a cost justification contradicting the nationwide rate-averaging philosophy, a principle firmly imbedded in the politics of the regulatory process if not in economic logic, and one that had served the common carriers well for years. Was it worth abandoning a favorable and familiar regulatory philosophy to justify a competitive response to one very small company? AT&T decided against an "exception tariff" on this route pending further study of the implications of such a decision.

The new specialized competitors were initially interested in the private-line market. Essentially, private lines are leased by the month while the more common switched-voice lines are available on a "dial up" basis, the line being paid for by the minute while in use.

In 1973 AT&T filed a new private-line tariff with reduced rates between the high-density areas most attractive to the new competitors, and increased private-line rates between the less vulnerable low-density areas. The total revenue effects on the company were neutral, but the rate-averaging principle had been abandoned. Rates on AT&T's switched-voice network, which was not subject to competition, remained unchanged, leading to later competitive interest by the specialized carriers whose facilities could provide both types of service. Thus, AT&T's return on assets would not be affected by the new tariff which the company believed should therefore be acceptable to the FCC.

The FCC began hearings on the tariff at the request of the specialized competitors, and after two years found the tariff to be unreasonable and in violation of the Communications Act. The AT&T replacement tariff submitted to the FCC, the Multi-Schedule Private Line (MPL) Tariff, was even more objectionable to the specialized competitors. Although allowed to go into effect in 1976, the new tariff was also investigated by the commission. An administrative law judge ruled in 1979 that the tariff should be rejected because of flawed cost-allocation procedures. However, for lack of an acceptable alternative, the FCC allowed the MPL tariff to remain in effect pending the development of adequate cost information.

Under the press of competition, AT&T decision making was becoming based on marginal-cost logic while it was still providing contradictory average cost data to the FCC as justification for new tariff filings. The contradiction was apparent to at least one of the FCC commissioners who was "sorely tempted to move the Commission to cancel the MPL tariff as a remedy for AT&T's patent and persistent failure to justify its rates and to comply with statutory and rule provisions and outstanding commission orders" (Brock 1987, 223).

As immediately recognized by Bell System management, the emergence of competition with the Bell System in the 1970s was not important in absolute terms; in 1975 Bell's share of the U.S. telephone market, limited by potential antitrust action, was 83.9% and its share of the private-line market was 96.7%. The specialized competitors' 3.3% of the private-line revenues were $35 million, which was only 0.1% of total U.S. telephone sales, and Bell management was well aware of the many ways by which it could continue to delay the growth of significant competition.

While government policy favored competition in specialized private-line markets, there was never any intention that the new carriers should become competitors of AT&T in the switched-voice market. In 1973 MCI sought a ruling from the FCC that it be allowed to provide FX service, a composite private-line/local telephone service by which a customer making a local call could reach, say, a store or an airline reservation desk in another city.

After much litigation and a favorable interpretation from an official at the FCC, the application was approved, and AT&T was required to provide interconnections. MCI, looking more and more like a competitor in the switched-voice market, entered in an indirect way in 1975 by filing a tariff for the Execunet service, a shared private-line concept. To the customer, the service was the same as the switched-voice service provided by AT&T, but the terms were different, combining lower rates with a minimum monthly charge.

AT&T complained to the FCC that the Execunet service constituted illegal competition with long distance service, and the FCC ordered MCI to discontinue the service. MCI appealed and was eventually successful, the court ruling in 1977 that the FCC had not shown that the public interest was best served by a monopoly in long distance telephone communications. AT&T attempted to deny interconnection services to Execunet customers, but this action was successfully challenged by MCI in the appeals court. The Supreme Court denied a stay requested by both AT&T and the FCC, letting stand the decision of the appeals court that local facilities must be provided to MCI. The policy of a national switched-voice telephone monopoly in the United States was being changed by the Supreme Court.

A new problem began to grow in importance: the proper charge for AT&T to levy on the other common carriers (OCCs), MCI, Sprint and the smaller companies, who invariably used Bell System local facilities to reach their customers. AT&T filed the ENFIA tariff which included much higher rates for local lines connected to Execunet-type services. As noted earlier, the Bell System had been subsidizing local telephone service through "Separations and Settlements" procedures, and believed that the OCCs should do likewise as competitors of AT&T.

The OCCs, with limited resources and lacking the 100-year-old national system viewpoint of AT&T, argued that the actual cost of local interconnect services was much less. MCI, stating that the ENFIA tariff was in violation of the court's ruling, complained to the FCC, which convened negotiations between the parties to avoid more litigation. The negotiations resulted in a compromise on the ENFIA tariff in 1978. By 1979 MCI's revenues were $43 million, the total revenues of the not-so-specialized carriers being about $50 million. The agreement specified that the OCCs charge per minute would be 35% of what AT&T was paying as its share of local costs, as long as local industry revenues were under $110 million per year. The rate would increase to 45% up to $375 million in revenues earned by the OCCs, and become 55% of AT&T's rate when OCC revenues reached $375 million. The constantly looming possibility of an antitrust suit may have been a factor in AT&T's agreeing to compromise on the ENFIA tariff. The continued existence of the OCCs created the impression of competition without actually doing very much harm to AT&T (Henck and Strassburg 1981, 181).

The OCCs thus had several cost advantages over their far larger competitor: they served only those routes they found profitable and they supplemented their low-cost long distance microwave facilities with AT&T's local telephone facilities at more-favorable rates. Thanks to the decisions of the Supreme Court, the specialized competitors were not limited to the relatively small private-line business; they were potential long-term players in the entire U.S. telecommunications market.

While the 1960s saw the emergence of competition with the Bell System in the provision of long distance communications services, the 1970s was a decade of growing competition in the terminal equipment market. The Hush-A-Phone and Carterphone cases, described earlier, set precedents for customer-owned attachments to Bell System telephones. AT&T responded by requiring protective connection attachments (PCAs) that would prevent harm to the telephone system, and by instituting a program of standardization and specifications for authorized telephone equipment.

The FCC established a joint federal-state board in 1972 to consider customer-owned network control signaling units and connecting arrangements, a significant step in the direction of permitting competition in the terminal equipment market. FCC policy was moving towards the elimination of Western Electric's monopoly position in favor of a competitive market for telephone equipment. At the same time, some of the state regulatory commissions, perhaps at the urging of the Bell operating companies, were moving in the opposite direction. Through the mechanism of residual pricing, local residential customers were being subsidized by favorable cost allocations at the expense of business customers using specialized services, and these subsidies were being threatened by the FCC's actions.

Competition in the business terminal equipment market implied lower interstate revenues and therefore might reduce the cross-subsidization of local telephone service. The North Carolina Utilities Commission decided to issue a rule prohibiting customer-furnished telephone equipment in the state. The FCC was asked to intervene by the North American Telephone Association, a group of independent telephone equipment manufacturers founded by Tom Carter of the Carterphone Company. The request was filed on behalf of Telerent Leasing, a North Carolina firm.

The FCC ruled that state commissions could take no action inconsistent with previous FCC rulings, thus preempting state authority over telephone interconnection matters. The North Carolina Commission, as well as the National Association of Regulatory Utility Commissions (NARUC), appealed the FCC ruling. The arguments against the FCC position centered on the responsibility of the telephone system to provide access to fire, health, safety and business facilities to everyone, and that the provision of this universal service as well as the reliability of the nationwide telephone system could not be guaranteed if "economic experiments" were permitted. Many state regulators viewed the national telephone monopoly as a "toll road" maintained for the benefit of all, but uneconomic to operate if anyone who wished could bypass the "toll booth" (Danielsen and Kamerschen 1986, 10–11).

In 1974 NARUC proposed legislation amending the Communications Act of 1934 to include the goal of universal service, which would require subsidization of telephone costs in rural areas, implying a strengthening of the interstate telecommunications monopoly. In 1976 the U.S. Court of Appeals in Richmond upheld the FCC decision to permit the connection of customer-owned telephone equipment, and subsequent appeals to the Supreme Court were rejected.

By 1976 The FCC was moving towards the eventual approval of customer-owned PBX and key systems, providing they were certified as meeting technical standards. A PBX, or private branch exchange, is a switchboard on the customer's premises that connects all of the telephones at the site to one or more outgoing lines. Modern PBXs, reducing the need for costly outside lines, are sophisticated electronic devices. A key system permits selecting an outgoing line by pressing a button on the telephone set.

By the late 1960s there was concern that the voice-oriented Bell telephone network might not be responsive to future demands for information and computer data transmission services, and that the telephone companies might have a competitive advantage over the data processing industry in the new markets. In 1966 the FCC began a study intended to clarify its policy in these areas. An important regulatory problem was that of avoiding the commingling of costs and revenues from the regulated communications business and the unregulated data processing activities.

The FCC issued a ruling in 1971 that was intended to separate the service offerings, but technological change was making it increasingly difficult to maintain the distinction between communication and data processing services. By the late 1970s any individual or business could purchase an inexpensive off-the-shelf modem and connect a terminal or personal computer to another computer anywhere through a telephone line ostensibly installed for voice communication purposes. The series of FCC rulings in the 1970s, backed up by court decisions, essentially deregulated the terminal equipment business, and the Execunet court decision, in spite of the FCC's active opposition, seemed to open up the national switched-voice market to competition as well.

In 1982, while the FCC was struggling with a system of access charges to replace the temporary ENFIA tariff, the Justice Department and AT&T surprised everyone by announcing agreement on a modified antitrust settlement under which the Bell operating companies were to be divested into separate corporate entities within two years. The AT&T company would retain 30% of the system assets, and seven new regional holding companies would each receive 10% of the assets.

The problem of local access charges in the more complex new industry structure became acute, and a reversal of the universal service philosophy was becoming evident. Rather than subsidize the local subscriber by allocating a portion of long distance revenues to the cost of connecting him to the system, as had been done in the past, the new reasoning was that the local subscriber was responsible for the cost of the local loop and should pay an end-user access charge. This idea is both consistent with economic logic and unavoidable as the telephone system became more competitive and the interstate revenues began to match costs more closely.

However, the political effects of sudden large nationwide increases in all residential telephone bills drew congressional interest and resulted in a phased, compromise solution to the access charge problem. The institution of access charges by the FCC, supported in the inevitable court case, was followed by AT&T filing a comprehensive tariff decreasing most long distance rates, which went into effect in 1984. Years of gradually increasing economic pressures on an outmoded telecommunications market structure and regulatory system were, at least partially, released in the sudden earthquake of the divestiture agreement of 1982.

CHAPTER 4 _____

The Telecommunications Industry Since 1984

AT&T had successfully fought antitrust litigation for almost a century, arguing that, as a natural monopoly, it was properly treated as a fully regulated public utility and that it was therefore immune from prosecution under antitrust legislation. The Department of Justice had filed an antitrust complaint against AT&T in 1949, but the company was successfully defended by the FCC, which argued that it had effective control over all aspects of AT&T's pricing (Henck and Strassburg 1988, 190–191).

By the early 1970s, the emergence of economically viable competitors, the prospect of rapid technological advances under competition and the growing evidence that much of the telecommunications industry did not have the characteristics of a natural monopoly, had all weakened AT&T's case.

The final series of events leading up to the divestiture of the Bell operating companies (BOCs) from AT&T began on November 20, 1974, with the filing of an antitrust suit by the Department of Justice, many private lawsuits alleging illegal business practices having already been filed against the company. The Justice Department charged that AT&T used its market dominance and control of local service "bottlenecks" to suppress competition and increase its monopoly power. Justice sought the divestiture of the BOCs and Western Electric from AT&T.

AT&T responded to the federal suit with its traditional argument that it was immune from antitrust suits because it was pervasively regulated at the state and federal levels, and requested that the suit be dismissed. The suit was not dismissed, and preparations for trial began. The Justice

Department, hampered by staff turnover and weak support from the Carter administration made slow progress preparing for trial (Henck and Strassburg 1988, 219). Judge Joseph C. Waddy, who was originally assigned to the case, died in 1976 and was replaced by Judge Harold Greene.

At this time the Department of Justice had been litigating a massive antitrust case against IBM for nine years, with little progress. The IBM suit was eventually dropped, but Judge Greene publicly committed himself to efficient management of the case against AT&T being tried in his court. In September 1978, the court, once again declining to dismiss the suit, ordered the parties to prepare for trial, setting deadlines in order to accelerate the pace of the litigation. During the ensuing several years of pretrial preparation, the Justice Department revised its suit, seeking only the divestiture of the Bell operating companies. AT&T declined to pursue settlement opportunities, possibly believing that it would receive sympathetic treatment from the incoming Reagan administration. Another factor in AT&T's decision was the large number of private lawsuits filed against the company, 49 by 1979, which had been encouraged by the Justice Department suit (Brock 1981, 295).

William Baxter, the newly appointed Assistant Attorney General for Antitrust, was against government intervention in business in principle, but he strongly favored the breakup of the Bell System, and he took active charge of the government's case. Baxter had previously been associated with one of AT&T's competitors and he was familiar with the telecommunications industry. Baxter's views were widely publicized, dashing AT&T's hopes for sympathetic treatment from the Reagan administration (Henck and Strassburg 1988, 224).

The trial began on January 15, 1981, but the parties, negotiating on the side, quickly announced a tentative settlement, and the trial was halted. The parties not being able to agree on the terms of a formal settlement, the trial resumed in March under the energetic prodding of Judge Greene. The government presentation was completed in August and was immediately followed by a request for dismissal by AT&T. The court declined to dismiss the case, stating: "The testimony and documentary evidence adduced by the Government demonstrate that the Bell System has violated the antitrust laws in a number of ways over a lengthy period of time . . . the evidence sustains the government's basic contentions, and the burden is on the defendants to refute the factual showings made in the government's case in chief" (Henck and Strassburg 1988, 226).

This statement by Judge Greene prior to AT&T's presentation of its case was ominous, because AT&T really was not disputing the facts of the case, and if it lost the case it would face numerous private triple-damage suits. Later statements by AT&T management indicated that the potential legal problems facing the company were so great that some sort of agreement breaking up AT&T along the lines sought by the Justice Department was inevitable.

AT&T resumed secret negotiations with the Justice Department in late 1981 while the case continued to be tried in court. On January 7, 1982, prior to the completion of AT&T's court presentation, the parties reached a settlement. AT&T would give up the local exchange facilities of the Bell operating companies, retaining about 30% of the assets of the Bell System. The Bell operating companies would be restricted to local-exchange common-carrier operations, and were to give up ownership of customer premises equipment (CPE) as well as the Yellow Page business. The Justice Department released AT&T from its 1956 consent decree which had prohibited AT&T from serving unregulated markets. An immense amount of effort was then expended on the division of Bell System assets between AT&T and the BOCs, and the actual divestiture, which began on January 1, 1984, was completed in September 1984.

AT&T management believed that, since it had given up the local exchange facilities that were regulated common-carrier monopolies, and the restrictions of the 1956 consent decree were removed, it was now free to "compete on an equal footing in all markets" (Faulhaber 1987, 84). Justice apparently agreed with this view, believing that the need to regulate AT&T would greatly diminish in the future.

AT&T and Justice interpreted the settlement as a modification of the 1956 consent decree, and requested that the responsibility for the decree be moved from the federal court in Newark to Judge Greene's court. It was supposed that Judge Greene would simply sign the modification of the 1956 consent decree thereby ending his participation in the case. Instead, Judge Greene invoked the Tunney Act, which specified a court inquiry to determine if the modified consent decree was in the public interest.

In 1982 the Senate Subcommittee on Communications, Science and Transportation held hearings on the settlement, and William Baxter testified that he believed the new AT&T markets to be "workably competitive." Baxter, in line with the settlement, continued to stress his hopes for the deregulation of AT&T. However, the FCC, while studying the matter, continued to treat AT&T as though it were still a regulated utility. It became clear that, while AT&T had reached agreement with the Justice Department, the settlement had no effect on the federal and state regulators that continued to possess the same statutory powers. At this time, legislation (HR 5158) was introduced in Congress that would have had the effect of increasing the regulation of AT&T and strengthening the monopoly power of the Bell operating companies. After intense lobbying by AT&T, the legislation failed to pass.

The possibility of AT&T being able to cross-subsidize, or use the profits earned in one market to support predatory pricing in another, had been of concern all through the 1970s. Although the divestiture settlement seemed to have solved the cross-subsidization problem, it did not solve the problem of AT&T's enormous market power in long distance, switching and transmission systems. AT&T, by far the dominant force in its markets, appeared to be a *de facto* monopolist. AT&T's competitors sought to obtain market advantages by claiming that continuing regulation of AT&T was needed, and the regulators themselves were not keen on eliminating their own jobs. The Bell operating companies argued that they were left with only the unprofitable local telephone business and therefore might not be economically viable. Finally, the adverse political effects of the divestiture were being felt by the state regulators, who faced the prospect of approving substantial local rate increases due to the loss of the long distance subsidies from AT&T.

Judge Greene, in this complex climate, maintained control of the divestiture case under the provisions of the Tunney Act. He described modifications to the settlement that he believed would be consistent with the Tunney Act, including permitting local Bell operating companies, whose viability as local common carriers was not clear, to sell new customer-premises equipment (CPE). Judge Greene's modifications came to be known as the Modified Final Judgment (MFJ). As mentioned previously, AT&T was to continue to own all existing, or imbedded, CPE. Judge Greene also left the lucrative Yellow Page distribution business in the hands of the operating companies.

Finally, in a decision that will probably have profound effects on technological progress in information systems in coming years, the judge listened to the pleas of the newspaper industry that the local telephone operating companies should be restricted from providing electronic information services that might compete with newspapers. Under the modified settlement, a Bell operating company could petition for a waiver of the line-of-business restriction if the proposed new business was less than 10% of existing BOC revenue, and if competition would not be harmed by entry into the business. By early 1986 Judge Greene had received 86 requests for waivers from the Bell operating companies and had granted 56, and there will probably be many more waivers in years to come.

The parties to the settlement accepted Judge Greene's modifications, the decree was entered, and very complex negotiations on financial and other matters ensued. Soon after, a committee of Bell operating company presidents proposed the reorganization of the 22 Bell subsidiaries into seven regional holding companies, each to own 10% of the former Bell System's assets, and the new AT&T to own 30%. After much wrangling over financial details, this proposal was put into effect. When the divestiture took place in 1984, AT&T issued shares in the seven new telephone holding companies to its 3 million shareholders. AT&T had lost most uses of its name and $87 billion in assets, but retained Western Electric, Bell Laboratories and its long distance business.

The new regional holding companies were promptly christened the "Baby Bells," but they were actually enormous companies with annual revenues ranging from $9 billion to $14 billion. They chose new corporate names, which were sometimes modified forms of the old Bell name, and sometimes entirely new names signifying their new beginnings. The New York and New England companies became Nynex; the Great Lakes companies became American Information Technologies Corp. (Ameritech); the companies in the northwest became US West; the California and Nevada companies became Pacific Telesis Group (Pactel); the mid-Atlantic companies became Bell Atlantic Corp.; the southeastern companies became BellSouth and the southwestern companies became Southwestern Bell Corp. AT&T was permitted to retain the use of the Bell Laboratories name and to use the Bell name in foreign operations.

After intensive negotiations over the boundaries of interstate long distance markets, Local Access and Transport Areas (LATAs) to be served by the Bell operating companies were carefully defined, and independent, that is, non-Bell, telephone companies became associated with adjoining LATAs. Intercity calls within the same LATA are serviced by the local operating company rather than by AT&T, and so the determination of LATA boundaries was of great economic importance, shifting some service previously defined as interstate long distance to the Bell operating companies and independent companies.

As mentioned above, many state regulators were unhappy with Judge Greene's modified final judgment, believing it to be an unlawful preemption of the regulatory authority of the states (Henck and Strassburg 1988, 234). It was thought that local telephone rates would have to increase substantially to cover the loss of cross-subsidization. In 1983 the Supreme

Court, responding to a complaint by officials of 13 states, affirmed Judge Greene's modified final judgment by a vote of six to three.

Many observers were concerned about Judge Greene's expanding role in the reorganization of the telecommunications industry: "By the end of summer 1982, Judge Greene had transformed a radical but carefully considered experiment in industrial economics into a radical crapshoot with one of America's essential infrastructure industries as the ante" (Faulhaber 1987, 101). An attempt to introduce legislation in Congress that would have removed Judge Greene from the telecommunications scene failed, largely due to opposition from the newspaper industry.

In spite of dire predictions concerning the effects of the decisions made by Judge Greene, which many thought extended too far into matters that should have been left to federal and state regulators, telecommunications services continued to be provided as usual throughout the United States as the divestiture took place. There were some problems while adjusting to the telecommunications reorganization: "A recent survey of 1,700 companies by the National Conference Board indicates widespread displeasure with the situation as a whole. . . . Over four-fifths of those surveyed reported that they believed service has deteriorated" (Danielsen and Kamerschen 1986, 13). However, the service problems were transitional, disappearing as companies and customers became familiar with the new system. Legal battles between AT&T and its former subsidiary companies over divestiture details are expected to continue behind the scene for years (Henck and Strassburg 1988, 246).

By 1988 the mix of services provided by the long distance market in the United States was composed of about 49% traditional residential and business long distance calls, 11% WATS services, 11% 800 services, 11% international services, 9% private-line services and about 9% credit card/Pay Phone/900 services (*Wall Street Journal*, January 17, 1990).

AT&T's share of the long distance market, which was 90% in 1984, had fallen to 70% by 1988. At that time MCI held 12% of the market with US Sprint in third place holding 8%. Metromedia/ITT, Telecom USA and Allnet each held slightly over 1% of the long distance market, the remaining 6.5% shared by all others. While AT&T still held a commanding position in the long distance market, it was clear that in order to remain competitive with leaner rivals such as MCI, substantial cost reductions were necessary.

Robert Allen, taking over as CEO of AT&T on April 19, 1988, believed that the most entrenched corporate culture in America must finally change.

The Company could no longer operate as the regulated monopoly it once was, satisfied with predictable price increases, complacent about competition, uncritical of costs. To compete in fast-changing markets—for that is where divestiture has taken all AT&T's businesses—the phone company of old had to learn to get aggressive, to take chances, and above all, to move quickly. (*Fortune*, June 19, 1989, 59).

In a few clear words Allen provided the justification for the deregulation of much of the telecommunications industry in the United States. Allen continues: "We came out of a business that had a single culture. It was very paternalistic, thorough, slow-moving and exceptionally proficient in accomplishing its mission, but in a static environment."

In a dramatic move, AT&T quickly wrote down $6.7 billion in obsolete transmission equipment, resulting in a $1.67 billion loss, the first such loss in 104 years. Obsolete transmission equipment was no longer legally entitled to a "fair rate-of-return" as it was under rate-of-return regulation, and good business sense dictated writing it off immediately. By 1989 AT&T was streamlining operations by automating factories and back-office facilities and by reorganizing into smaller groups focused on individual businesses.

AT&T announced improvements in its pension and early retirement plans intended to induce about one-third of its eligible managers to retire, at a potential savings to the company of $450 million in the following year. AT&T announced that it had cut its total employment by 25,000 in 1989 and expected a further reduction of 8,500 in 1990. "We're on a drive to right-size this place to be competitive" said Harold Burlingame, vice president of human resources. "Our managers are quickly learning that they must focus on costs, customers and competitors" (*Wall Street Journal*, December 11, 1989). Cost-cutting moves since 1984 had reduced AT&T employment by 90,000 to a total of 280,000 by the beginning of 1990, with an ultimate target employment of about 250,000.

The Paradyne Corporation, a manufacturer of data communications equipment, was acquired by AT&T for $250 million in 1989, indicating the Company's new receptive attitude towards externally developed technology. The unsuccessful cooperation with the European computer firm, Olivetti, was terminated, and new approaches to the computer business were planned. In late 1990 AT&T announced an intention to acquire NCR Corp. in order to take advantage of that company's proficiency in computer networking.

The Bell operating companies, now organized in seven regional holding companies, are also facing the exigencies of competition, and they are reducing employment and streamlining operations. Since local telephone exchange operations are still regulated, revenue growth from basic services is usually governed by economic and population growth, which vary from region to region. Conventional local telephone service is a passive, unexciting market, and many feared for the economic viability of the Bell operating companies under politically sensitive regulation by state commissions. While those regional Bell companies in high-growth areas benefited from the installation of new telephone lines in the 1980s, most found that the fastest domestic revenue growth businesses were network access services and business telephone traffic. However, by the end of the 1980s, most of the Bell operating companies had become aware of astonishing new business opportunities, described later, that promise to transform their markets.

In addition to cost cutting and a surge in innovative activity, there is now interest in better management on the part of the Bell operating companies. Raymond Smith, chair and chief executive officer of Bell Atlantic Corporation, stated "We want to compete globally. But before we can do that, we have to become competitive internally" (*Wall Street Journal*, July 12, 1989). Corporate departments providing services such as data processing were to charge for their services and use the resulting revenues to cover their expenses. Ten departments in Bell Atlantic now sell their services to other departments in the company, in competition with possible outside vendors. While inconvenient at times because of the necessary formalities, the program successfully emphasizes the efficient provision of those internal company services that are the most valuable and the elimination of low-value activities.

The former cross-subsidization policy of the regulated Bell System has left a legacy of problems that will only slowly be corrected. There was a great deal of controversy over the long distance access charges that basic telephone customers would pay after the divestiture. The final compromise figure was lower than cost, and so the long distance companies, AT&T and its competitors, were still required to subsidize local telephone service, although the subsidies were smaller than they were in the past and scheduled to decrease in the future. MCI and US Sprint successfully argued that their payments to the local companies should be less than AT&T's, and a sliding scale was worked out. The question of access charges is both important and very technical, and is given detailed consideration in a later chapter.

Large organizations using sophisticated telecommunications and information systems have been the greatest beneficiaries of the growing competition in telecommunications that resulted from the breakup of the

Bell System. The economic benefits conferred on society by the introduction of a great variety of low-cost innovative telecommunications services into business, nonprofit and governmental organizations are usually indirect and difficult to measure but are undoubtedly enormous. The timeliness, adaptability and quality provided by modern business communications systems are essential for prompt responses to changing market conditions and the efficient management of assets in a fast-moving, competitive world.

The new companies manufacturing, distributing and maintaining customer-premises equipment are also benefiting from competition, and the traditional companies in this industry are finding it necessary to become efficient and responsive to customer needs in order to prosper. It was initially believed that when the CPE business was opened to competition much of the business would be lost to foreign competitors. For a few years a good deal of business was lost, but the U.S. companies now producing PBX and other customer systems are competitive in all respects, and the United States is a net exporter of these products.

Residential customers, long accustomed to the paternalistic practices of the Bell System, were caught by surprise by the changes in billing and equipment maintenance procedures that went into effect as the divestiture took place, but as the new arrangements became familiar the initial confusion subsided. As predicted, most small business and residential subscribers saw increased charges for conventional local telephone services. While they could now own the telephone equipment on their premises, and many bought the handsets previously leased from the telephone company, residential customers did not rush in large numbers to buy the new equipment offered in designer styles and colors.

Residential telephone subscribers are benefiting from lower long distance rates and a variety of innovations in service, mostly due to competition between AT&T and its two largest competitors, MCI and US Sprint. AT&T's basic long distance rates had dropped 30% by 1987, and of course its competitors had to match the new rates. Both AT&T and its competitors are offering new long distance service plans such as "Reach Out America," designed to be attractive to various customer classes. In most cases, the savings on residential long distance service are probably less than the rate increases for local service, increased access charges and the costs of occasional maintenance problems.

Nevertheless, economists would argue that the overarching benefits provided by competition in the telecommunications industry are more than worth the higher costs of residential service, which existed previously but were invisible to residential customers, having been subsidized by others.

Local telephone technology was stagnating prior to the period of divestiture, and innovative new products and services were slowly introduced into local service, if at all, in the regulated monopoly setting. At this time of writing, large numbers of rural telephone customers still do not have touch-tone service. The key to lower local telephone rates, or, of equal benefit to the consumer, improved services at current rates, is innovation. Many of the new regional holding companies are now aggressively pursuing innovations in local telecommunications services that are likely to revolutionize the residential telephone business in the 1990s and beyond. Unfortunately, as described later, many of these innovations are being introduced into unregulated overseas markets rather than in the regulated local markets at home. The major benefits to residential customers of the era of competition will be seen in the future, when residential customers are likely to have a startlingly wide variety of economical telecommunications services to choose from.

At present, the extent to which the regional telecommunication monopolies may exploit new technologies by entering new businesses or providing new services is limited by the modified final judgment entered by Judge Greene. The Bell operating companies are currently permitted to provide exchange and exchange access telephone service, may publish and distribute Yellow Page directories, and may manufacture and sell new customer premises equipment. Consequently, many of the regional Bell companies are actively seeking new businesses overseas, where their operations are not regulated and potential profits are larger. The possibility of these companies commingling regulated and unregulated revenues, as well as the likelihood of new problems in allocating the common costs of regulated domestic businesses and unregulated foreign businesses, indicates a proliferation of new problems for regulators in the future. At present, the FCC is relying on standard accounting procedures to keep regulated and unregulated revenues and costs separate.

As state and federal regulators have discovered over the years, and as Judge Greene has undoubtedly discovered all over again, the ongoing implementation of regulation is an extraordinarily difficult task, and may be impossible to accomplish in detail considering the enormous size and operating complexity of each of the seven regional holding companies and their operating subsidiaries.

Some of the regional monopolies are introducing data transmission and information services. For example, Bell of Pennsylvania, a Bell Atlantic company, has introduced its Gateway service, advertised as a

"Supermarket for your computer. Plug it into a modem and open its eyes to a whole range of convenient, ready-to-use information services." Gateway was offered at a price of $5 per month plus the cost of each telephone call.

Prodigy Services Co., an on-line electronic information service, transferred its data storage facilities in the northeast to Nynex, and is working on similar arrangements with other phone companies across the nation. Nynex will operate and maintain the information delivery and storage systems. The agreement will allow Prodigy, a joint venture between IBM and Sears, to reduce costly capital expenditures associated with computer networks so that the company can concentrate on marketing. The telephone companies stand to gain increased traffic. (*Wall Street Journal*, December 7, 1989). Prodigy offers access to hundreds of data bases and services, including home banking, for a monthly base fee of about $10 in 1989. Prodigy customers must have a personal computer and modem to gain access to the system, but less than 20% of American households own such equipment. Still, the First Interstate Bank of Denver called the newly introduced Prodigy home banking service "the most successful new product launch this bank has had in several years." IBM and Sears have invested upwards of $600 million in the Prodigy venture.

Pacific Telesis Group and other regional Bell companies are active participants in cellular communications markets in the United States and in Europe. The number of portable telephones in the nation had grown to 2.5 million by 1989, with 10% of them located in California: "Fanatics are taking their new toys into restaurants—much to the dismay of diners and restaurateurs," reports the business press.

In late 1989 BellSouth was engaged in a bidding war with McCaw Cellular Communications Inc. over the control of Lin Broadcasting Corporation, another cellular communications company. McCaw, at that time the largest cellular telephone company in the United States, provided service to 127 markets.

McCaw and similar cellular telephone companies are potential future competitors with the regional Bell companies in the basic telephone service market, since only the greater expense of the portable telephone keeps it from replacing the household wired telephone handset. But this economic barrier is probably already breaking down in affluent households, where cellular phones purchased for business applications are often used for private purposes when convenient.

The concept of the local telephone company as a natural monopoly due to its paired-wire analog technology, a concept at the heart of the Bell System divestiture agreement, is breaking down as technology advances. Of course, once it becomes clear that local telephone companies are no longer natural monopolies, the case for permitting competition in basic local telephone services just as in the long distance market becomes strong.

Many of the regional Bell companies are interested in foreign telephone operations and equipment sales, perhaps as experiments with possible technological and business lessons for future U.S. operations, but certainly for the prospect of unregulated profits. At the beginning of 1990, the most aggressive of the foreign venturers was Pacific Telesis Group, which was drawing only 5% of its revenues from its overseas businesses. But foreign telecommunications operations are expected to grow in importance for most of the regional Bell companies as well as for AT&T. Only Ameritech, the midwest holding company, after an aborted attempt at securing foreign business, sees greater risks overseas and plans to concentrate on exploiting opportunities at home.

Pacific Telesis, after winning a competition with five of the other regional Bell companies, is a member of a consortium that received a license to build the first cellular telephone system in (West) Germany. This affluent, densely populated market is expected to be very profitable in the future.

In addition to its cellular plans, Pacific Telesis has invested heavily in foreign wireless communications, cable television and information services. Pacific Telesis, as a part of a consortium, is competing for a license from the British government to build a "personal communications" network using innovative radio technology to provide digital communications and information services. Pacific Telesis also owns interests in three British cable television companies. Pactel is part owner of a Japanese long distance telephone network and, with permission from Judge Greene, is participating in the laying of a fiber-optic cable between the United States and Japan.

US West, as leader of a consortium of seven foreign companies and a Soviet ministry, plans to lay a fiber-optic cable across the Soviet Union. The 12,000-mile cable would be the longest in the world and would link Europe with Japan, completing the 'round-the-world fiber-optic network.

US West signed an agreement with the government of Hungary to build a mobile cellular telephone system in Budapest (*Wall Street Journal*, December 5, 1989), the first such telephone network to be constructed in Eastern Europe. The existing telephone network in Hungary is obsolete, and the demand for telephone service greatly exceeds the supply. Hungarians are expected to use cellular telephones for both mobile communications and basic home service. It is likely to be cheaper for Eastern European countries, with antiquated telephone systems, to provide telephone service with cellular networks rather than replace or expand conventional copper-wire networks. US West will own 49% of the cellular network and Magyar Posta, the Hungarian postal telegraph and telephone system, will own the remainder. US West also owns interests in seven cable television franchises in Britain, and part of Lyonnaise Communications, which controls 13 cable television networks in France.

Nynex is planning a showcase telephone system for Gibraltar, and BellSouth is active in France and South America. Bell Atlantic is marketing consulting and other services in West Germany, Austria and Switzerland.

AT&T is also actively seeking business opportunities in foreign markets for telecommunications services and systems. AT&T received a $7 million order for digital switching equipment from Poland's state telephone agency. The switching system, with a capacity of 7,710 trunk lines, will be installed in Warsaw.

In addition to foreign operations, the regional Bell companies are aggressively pursuing new domestic businesses in permissible areas. US West was the first of the regional Bell companies to offer an electronic mail service that takes the place of a telephone answering machine. Customers ordering the service record a greeting for callers that is stored at the telephone company central office. The service allows callers to leave messages when the line is busy, a feature not provided by answering machines. When convenient, the customer calls an "electronic mailbox" containing messages by dialing an access number and a personal code number. By late 1989 several other companies were planning on offering similar electronic mail services.

The electronic mailbox is, of course, a substitute for the answering machine, which is believed to be in use in 24% of American households. According to the Electronic Industry Association, factory sales of answering machines totaled $545 million in 1988. The prospective $5 per month cost of the answering service makes it more expensive than the one-time cost of purchasing an answering machine, but the latter requires tapes and maintenance and has limited features. A Bell of Pennsylvania representative thinks that 41% of the consumers now using answering machines would be interested in the new service, which would involve more local telephone traffic.

The interesting aspect of this lively competition between alternative ways to record telephone messages is that consumers benefit significantly by being able freely to make this kind of choice. If the more expensive electronic mailbox provides features considered valuable by a particular consumer, that consumer is free to sign up for the service. If enough consumers sign up, the service will be continued and probably be extended to other areas. Otherwise, the service will be discontinued. In any case, the existence of several types of answering service products provides the competition necessary to keep prices low, and free entry into the market implies that when more-economical alternatives are developed, they will also become available.

Caller ID, a new service identifying the caller to the recipient of the call, has been proposed to the Public Utility Commission by Bell of Pennsylvania. There is significant controversy about this proposed service, which some argue may violate the telephone tapping provisions of the Electronic Surveillance Control Act. The privacy of communications has long been a central feature of telephone service and is protected by law. Ironically, the law was originally favored by AT&T as a way of preventing "foreign" connections from being made to Bell System equipment. Now that any terminal equipment meeting FCC specifications may be used by the consumer, the law tends to discourage the introduction of some new products by the regional Bell companies since some of the proposed new telephone services are seen by critics as permitting opportunities for violations of the right to privacy.

Facsimile machines, often with built-in telephones, have been standardized and have come into wide use in the 1980s, permitting the quick, economical transmission of the image of a typed page or photograph to any other compatible fax connected to the worldwide telephone system. Technological developments in this area combined with the growth of information services are a potential competitive threat to the newspaper and publishing industries.

The menu of new telecommunications products offered in the United States and overseas in the 1980s is rapidly expanding, and revenues are growing briskly. Annual revenues of AT&T and the Bell operating companies, which had exceeded $60 billion in 1980, continued to grow in the late 1980s, exceeding $106 billion in 1986 (U.S. Dept. of Commerce, Statistical Abstracts, 1989). However, under competitive pressures to achieve increased efficiencies, the total employment of AT&T and the former Bell operating companies decreased from a 1981 high of 945,000 to 742,000 in 1984, and continued decreasing in the late 1980s.

In spite of the dire predictions made at the time of the breakup of the Bell System, from the viewpoint of early 1991 the institution of competition in telecommunications markets has been a resounding success, and the industry structure established in the United States will probably form a model for other industrialized countries in coming years. Most interesting is the commercialization of new technologies such as fiber-optic cable, digital cellular telephone systems and wireless systems that may eliminate the natural monopoly aspects of the local telephone and cable TV networks and eventually release the forces of competition in these markets as well as in the long distance market.

An important argument against the breakup of AT&T and the primary argument of the Department of Defense was that an integrated national telephone system was inherently more reliable than the chaotic system likely to develop under competition.

On January 15, 1990, a software failure occurred in AT&T's network that shut down long distance traffic for nine hours. Where the old monolithic AT&T management might have been phlegmatic and aloof in the face of such an occurrence, the new AT&T management was abashed and apologetic. The AT&T breakdown was immediately publicized by MCI and US Sprint in their advertising, both companies suggesting that customers requiring reliable service should, of course, sign up with them or at least sign up with more than one long distance provider. Similar advice was offered earlier by US Sprint when AT&T suffered some service disruptions due to Hurricane Hugo. In fact, as explained in the business press that week (*Wall Street Journal*, January 17, 1990), most customers normally using AT&T service could have reached other long distance carriers during the breakdown by simply dialing their access codes.

The prompt exploitation of the AT&T failure by MCI and US Sprint in their advertising, AT&T having behaved in a similar fashion in the past, was a comforting indication that effective competition exists and that there is no need to worry, for now, about the long distance telecommunications industry becoming collusive.

In actuality, the existence of competition in long distance service offers the interested customer greater reliability through redundancy of providers as well as putting intense pressure on the market leader, AT&T, to operate as reliably as possible. There is no reason to believe that post-divestiture long distance telecommunications services are any less reliable than before, and opportunities provided by redundancy improve system reliability from the customer's standpoint.

The 1980s began with the breakup of the familiar Bell System, and the public suffered through a brief period of confusion and spotty service as the entire industry was reorganized. By the end of the 1980s the public had forgotten the old system, and service in general was more reliable and more economical than ever, while many innovative new products promising further dramatic changes in the telecommunications market were being developed and introduced.

CHAPTER 5

The Determinants of Telecommunications Market Structure

Market structure, the configuration of buyers and sellers in a market, plays a central role in both price determination and the rate of innovation in an industry, where the industry is defined as the sellers of the product. Innovative activity, in turn, affects the trajectory of market performance over long periods of time. Since a market is composed of both the buyers and the sellers of a product, its structure is, in part, determined by the number and sophistication of buyers as well as the number of sellers. As the number of buyers and sellers increases, competition tends to become more effective. There are numerous bids for available supplies of the product as well as numerous offers, and the likelihood of collusion among buyers or sellers diminishes.

The structure of the industry—the number and size distribution of firms selling in the market, the conditions of entry and the degree of concentration of business among firms—is generally formed by the technical conditions of producing and marketing the industry's products, by the characteristics of consumer demand and by the legal and regulatory environment in which the industry operates. Regulatory policy, when it exists, shapes industry structure and behavior, with the intention if not always the effect of bringing about socially desirable market performance.

Controversy arises over the precise criteria to use in evaluating economic performance. Regulatory officials, working within a circumscribed legal framework, tend to emphasize fairness at the expense of either static or dynamic economic efficiency. Economists use the public interest as a criterion, usually emphasizing the achievement of economic efficiency according to the comparative statics measures defining pure competition. Some economists, following Schumpeter (1942), attempt to apply more-sophisticated criteria, emphasizing a high rate of techno- logical progress and innovation as indicative of dynamic efficiency, even if the market structure does not satisfy static criteria for efficiency in the short run. In this chapter, alternative market structures are compared with respect to the likelihood of achieving economic efficiency in both the static and dynamic senses in the telecommunications industry.

The telecommunications market in the United States at this time of writing is a hybrid oligopolistic market in which a few large national telecommunications companies dominate the long distance businesses, with many small competitors occupying specialized product niches. The common-carrier operations of the largest firm in the interstate market, AT&T, are regulated by the FCC, but AT&T's competitors are not regulated, and much of this market is quite competitive. Regional monopolies provide local exchange service under state regulation and also sell or support an increasing variety of services and products under competitive conditions in the United States and overseas.

The availability of innovative bypass services, encouraged by the uneconomic regulation of rates and the high rate of technological progress in digital radio technologies, is reducing the traditional market power of the local telephone companies. Customer-premises equipment and certain specialized services are produced and sold under competitive conditions in the United States by the Bell operating companies, and the scope of these activities appears to be expanding rapidly. Cellular and other wireless networks, sometimes operated by subsidiaries of the Bell operating companies, are being introduced in local areas under FCC regulation. Cable TV networks are local monopolies, franchised by municipalities, with regulatory oversight provided by the FCC as in the case of broadcasting.

The existing market structure evolved from a history of federal and state regulation, a series of court decisions, the AT&T divestiture agreement with the Justice Department and the Modified Final Judgment (MFJ) of Judge Harold Greene, and not from a national telecommunica- tions plan or from a reaction to changing market conditions. It is very likely that it is simply a transitional structure that will continue to be modified in coming years in response to economic and technological opportunities, court decisions and changes in regulatory policy at the state and federal levels.

The future telecommunications market structure in the United States is likely to be characterized by increased competition in both local and long distance markets in the near term, with oligopolies eventually assuming control of both local and interstate markets in the long term as technologies become standardized and smaller companies find it more and more difficult to match the production and marketing efficiencies of the large firms.

Much of the present competitive activity in telecommunications is due to economic opportunities created by asymmetric regulation. These opportunities will eventually disappear as regulatory policy is revised and the market shakes out, and so will the specialized companies formed to take advantage of them. Regulation at the federal and state levels is likely to decrease as increasing customer telecommunications options and declining costs of providing service reduce the market power of the traditional telephone companies. It is likely that the traditional local paired-wire analog technology employed by the regional telephone monopolies will be challenged by fiber-optic cable, digital cellular and other wireless communications systems. The trend away from paired-wire analog systems is already evident in some European countries in which private firms employing new technologies are actively competing with the former national telecommunication monopolies.

An analysis of the evolving telecommunications market structure begins with a brief description of the theoretical model of perfect competition, not because an industry composed of numerous small firms is a realistic model of a modern telecommunications market but because this model serves as a theoretical ideal with which to compare the former, existing and likely future telecommunications market structures in the United States.

Competitive markets, by allowing free rein to individual initiative, provide incentives to individuals to produce those goods and services most desired by others while also providing incentives for the efficient use of limited human and material resources. Markets tend to be efficient when prices are determined by the forces of aggregate supply and demand rather than by the decisions of a few individuals or firms. Efficient markets tend to maximize collective economic welfare in the sense that voluntary exchange permits individuals to maximize their satisfaction in the economic dimension by freely choosing production and consumption decisions.

While these ideas may sound trite and not specifically relevant to the real world, one has only to consider the many documented instances of economic stagnation resulting from poorly conceived government regu-

lation or the economic misery existing in many socialist planned economies in order to understand their vital importance.

Competitive markets are characterized by: (1) large numbers of
price-taking buyers and sellers, (2) free entry into and exit from markets,
(3) relevant information freely available to all, and (4) all sellers offering
a standard or homogeneous product. The fulfillment of all four conditions
is rarely encountered in reality, but markets that come reasonably close
to meeting these conditions appear to perform quite well.

The recent theory of contestable markets (Baumol, Panzar, and Willig
1984) emphasizes that just the threat of entry by potential competitors is
usually sufficient to deter a dominant firm from charging a monopoly
price or exercising market power, even though the number of actual
sellers in the market is small. A market is said to be contestable if actual
and potential sellers have access to the same technology, buyers respond
quickly to price changes and sunk costs are not an obstruction to free
entry or exit.

Because production technology and the requirements for mass marketing are likely to dictate that only a few firms serve future telecommunications markets, the probability that such markets can remain workably
competitive through the mechanism of contestability is an important
consideration. Contestability implies that even a few companies in a
market will have incentives to introduce innovative products in competition with existing products in order to deter the competitive activities
of potential entrants, but there is no guarantee that the entrants will not
be successful. Contestability appears to be a stronger incentive for
competitive behavior when an industry is characterized by a high rate of
technological progress.

A fundamental characteristic of a competitive market is that the market
price tends to reflect the marginal cost of production. Marginal cost is
the cost of the final increment of production, and markets tend to be
competitive when the typical firm's marginal cost begins to rise at a level
of production that is small relative to the market demand. Rising marginal
cost as output expands forms an effective upper limit on the ultimate size
of the firms in the market because each firm maximizes profit by
producing up to the point where the cost of the last increment of product
is equal to the market price. A firm can expand production past the point
where marginal cost equals market price only by accepting a lower level
of profit. The essential role played by technology as a determinant of
market structure is that it governs the behavior of marginal cost as output
increases. In the long run, the existence of competitive entry tends to

reduce profits such that price approaches average cost at a point where price is also equal to marginal cost.

A market can remain competitive indefinitely if production technology and input cost considerations cause marginal cost to rise in such a way that the sizes of the profit-maximizing firms in the market are small relative to the total size of the market, thus ensuring that no one firm is large enough to affect the market price by its production decisions. On the other hand, if the production, distribution and marketing technologies are such that marginal cost falls with increasing output, one or a few very large firms will eventually come to control the market.

Competitive markets are characterized by the absence of sellers with market power, defined as sellers with the ability to increase profits by controlling price through production and marketing decisions. The possession of market power is not illegal, but the means used to acquire and maintain market power are restricted by law. Market power is often indicated by the existence of only a few sellers in the market, and is related to, but not necessarily determined by, the concentration ratio, often defined as the share of the market held by the four largest firms. Market power is also affected by factors such as product characteristics, availability of substitutes, technological progress, geography, and the passage of time.

Market power is reduced when the product possesses characteristics that are also provided by reasonably close substitutes. Market power in a geographical area is reduced to the extent that customers can conveniently shop for the same product in other areas, or the product is easily shipped to the customer's area. Market power tends to fall over time as substitutes for the product appear in response to above-average profits or as consumer investments that generate the demand for the product are replaced or as consumer preferences change. Market power is reduced as production cost falls with advancing technology. In the long run, the entry of firms attracted by above-average profits, if successful, tends to diminish market power by increasing customer options, causing the market price to approach the long run-average cost of production.

Competitive markets have many desirable attributes, including incentives tending to minimize production costs and maximize social welfare, the latter being defined as the sum of the net benefits received by consumers and producers participating in the market. Free entry, in particular, enables a new firm with access to a superior production technology or an improved product quickly to become a participant in the market to the benefit of the new firm's backers as well as to the public.

The usual description of a competitive market emphasizes the static approach to economic analysis, portraying price as the driving force and downplaying dynamic effects such as technological change, innovation and new forms of business arrangements. Schumpeter (1942) believed that innovation was the most important competitive force, with price playing a minor role. But the dynamic innovation process is difficult to represent using analytical models, and so it is often relegated to a secondary status in economic analysis. The "price cap" regulatory policy of the FCC announced in May 1988, described in a following chapter, is intended to provide incentives for innovation and is therefore an exception to the tendency to apply static economic criteria to regulatory problems.

Based on the brief history of the U.S. telecommunications market given in previous chapters, it is clear that the ongoing discovery of new technologies, products, production processes and business arrangements in response to the prospect of economic gain, and the introduction of these products, processes and business services at prices competitive with existing products, constitute the very essence of competitive behavior in that market. Technological advances and innovation are the driving competitive forces in modern telecommunications markets.

It is often the case that sellers in otherwise competitive markets are successful in differentiating their products, thereby permitting themselves some discretion to set their prices above the usual market price. This very common type of market behavior is called monopolistic competition and is not considered justification for government intervention, because consumers with cheaper alternatives evidently prefer the differentiated products and will pay a higher price for what they perceive is a superior package of services.

Such markets can become dominated by a few large firms if economies of marketing or production give larger firms a cost advantage and concentration increases in the industry. Oligopolistic firms employ modern marketing techniques to emphasize the desirability of their products while reducing price sensitivity on the part of their customers, and prices are usually higher than they would be in competitive markets as a result. Oligopolistic markets are characterized by few sellers, but are not believed to require formal government regulation beyond the existing body of law intended to protect the public from abuses arising from the possession of market power. The behavior of such industries is subject to the ongoing scrutiny of the Department of Justice as well as state attorneys general. The threat of antitrust action by the Department of Justice, such action usually acc   by private lawsuits seeking triple damages, is enough to retain at least the appearance of competition in an oligopolistic industry.

Oligopolistic industries, especially if they employ high technology, are likely to remain workably competitive over time if they are contestable and if the antitrust laws are vigorously applied. The informal cooperation sometimes observed in oligopolistic markets, usually in the form of tacit agreements not to compete in terms of price and to respect one another's markets, sometimes develops into an unstable relationship as the market matures, and may finally decay into competitive behavior, assuming that mergers are regulated. Contestability and the possibility of imports provide additional incentives encouraging competitive behavior.

Large modern firms, in order to maintain their strong market positions, engage in organized research and development activities and typically produce useful products of high quality that are perceived as good values by their customers. These large firms emphasize efficiency in production and distribution and require continuing high-volume sales, which implies continuing customer satisfaction. Competition in such markets tends to take the dynamic form in which new products are constantly being developed and marketed in competition with existing products, rather than the static price competition of economic theory.

The possession of the financial resources necessary for continuing investments in research and development and in new production and distribution facilities and the requirement to deter competitive entry tend gradually to decrease price over time in oligopolistic industries. In spite of the comments of observers such as Galbraith (1967), who are concerned about the market power of the large oligopolies, it is by no means clear that consumers are worse off being served by such firms under contestable conditions than they would be if served by many small high-cost competitive firms utilizing simpler production technologies.

A recent development in the United States tending to offset the power of oligopolies is the growth of imports and the rise of foreign-owned domestic plants manufacturing what had been foreign products several years earlier. Competition from foreign firms reduces the market power of domestic oligopolistic firms while increasing the variety of consumer product choices and spurring technological advances and innovation.

During the 1980s a number of domestic industries were deregulated in the belief that the public would be better served by relying on competition to provide market discipline. The jury is still out on the benefits of deregulation in industries such as the airlines, which have quickly organized themselves to obtain and exploit substantial market power in the regions they serve. In general, however, the public appears better served when government regulation is replaced by competition.

To summarize, the frequently stigmatized oligopoly market structure may be the only form that can approach dynamic economic efficiency in the telecommunications and other high-technology markets. While short-run prices are likely to be higher due to market power, the resulting above-average profits earned by very large firms provide the substantial sums necessary to support research and development activities. Competition takes the form of continuous introduction of innovative products that compete with both existing products and those of potential entrants, the latter attracted by the possibility of above-average profits.

The discussion now turns to the economic characteristics of natural monopoly markets. A market may be inherently noncompetitive for various technical reasons, often because the cost per unit of product declines with increasing scale of production beyond the point at which the entire market is served. For example, in a conventional telephone system the cost associated with the addition of a new customer is lower than the costs previously incurred to serve existing customers because much of the installed equipment is used in common by all customers. A new telephone company formed to compete with an existing company must duplicate the facilities of the existing company, and the cost of serving each customer is therefore much greater.

When one firm can satisfy the market demand for a product or service more efficiently than can several firms, the market is considered a natural monopoly. In contrast to the competitive model described above, firm size is not limited by rising marginal cost, and there is nothing to prevent one firm from expanding to a size large enough to serve the entire market. Such a situation is a mixed blessing for society. While such a firm is capable of satisfying the entire market demand at the lowest possible price, its owners may prefer to use their market power to control production and market price in the interest of increasing profits. Government regulation of such firms is intended to preserve the benefits of efficient production for the public while ensuring that the owners of the firm receive no more than a fair return on their investment. However, government regulators are sometimes more sympathetic to the monopolistic firm and its problems than they are to the problems of consumers.

The traditional telecommunications market structure in the United States consisted of a privately owned integrated national monopoly with government regulation of the subsidiary operating companies at both state and federal levels. AT&T had long argued that this type of market structure was socially optimal because the telecommunications industry was a natural monopoly and the national market could be most efficiently and reliably served by a single firm. Secondary arguments advanced by AT&T and its supporters over the years included: the economic efficiencies available due to integrated management; the superior response of the captive Western Electric Company, the system's manufacturing subsidiary; the necessity for national standardization to maintain high levels of system performance, reliability and safety; a lower cost of capital enjoyed by the subsidiary companies as a consequence of the financial strength of the entire system; and the strategic national-defense value of an integrated telecommunications network. As shown later, none of these supposed advantages of a national telecommunications monopoly is supported by the objective study of relevant data or by the events that have occurred since the divestiture of the Bell operating companies from AT&T in 1984.

Of primary interest in the present study, an important property of a socially optimal market structure is the provision of adequate incentives for process and product innovations. Despite constant reference to the renowned technological prowess of Bell Laboratories, history shows that innovative activity was not a strong point of the former national telephone monopoly. As described in Chapter 2, AT&T used the research capabilities of Bell Laboratories to monopolize telecommunications technology in order to protect its markets, and innovation in the Bell System was often a reluctant response to actual or potential competitive pressures arising through legal challenges.

Not so long ago telegrams were delivered by bicycle or other vehicle so as to prevent any contact between the Western Union system and the Bell System, although European telegraph and telephone systems had been integrated from the very beginning as a matter of common sense. While readers of a certain age may recall many similar anecdotes, the previous chapters provide more substantial support for the claim that the Bell System monopoly did not actively pursue and introduce innovative consumer products and services. Rather, arguments supporting the national telephone monopoly structure rested on the static economics of telephone service, and it is these arguments that are now studied in detail.

A natural monopoly is said to exist when economies of scale lower the cost per unit of service as output expands to include the entire market. Scale economies arise when the technical aspects of production are such that the rate of production may be increased without requiring proportionate increases in all inputs, thus lowering the average cost per unit produced, where average cost is defined as the total system cost divided by the rate of production.

While the above traditional definition of natural monopoly is satisfactory for many purposes, Baumol (1977) has shown that this definition is not sufficient for the precise treatment of multiple-product monopolies. The modern definition of a multiproduct natural monopoly makes use of the concept of subadditivity, which implies that every output combination is produced more cheaply by a single firm than by multiple firms, all employing the same technology.

The word *technology* refers to a specified relationship between the rates of production of one or more products and the rates of consumption of one or more inputs to production. This relationship is called a production function. Suppose that the market demand is Q per unit of time, and two firms supply the market, with firm A providing the positive fraction f of the market demand, or fQ, and firm B providing $(1 - f)Q$ of the market demand. Let total cost be a function of the rate of production using a given technology available to both firms, and denote this function $C(Q)$. Average cost is then $C(Q)/Q$. If the total cost incurred by a single firm to satisfy a given market demand, $C(Q)$, is less than the sum of the costs of two firms sharing the market, $C(fQ) + C[(1 - f)Q]$, then the costs are said to be locally subadditive and the market is a natural monopoly at that level of market demand. If costs are subadditive at all levels of market demand, then costs are globally subadditive and the market is a natural monopoly at all levels of demand. The distinction between locally subadditive and globally subadditive is important when, for example, a market is a natural monopoly at a low level of market demand, but costs are not globally subadditive and the market becomes competitive as demand increases. These definitions assume a static technology. A market that is globally subadditive under a specific production technology may not be so when the technology changes.

These definitions have important implications for the price paid by consumers. Market price will tend to approach average cost in the long run if regulators or competition ensure that no firm earns more than a fair rate of return. Average cost is lower when a single firm serves a market when costs are subadditive, that is, when a market is a natural monopoly. Consequently, in static theory, consumers are better served by a regulated monopoly when costs are subadditive, and this is the intellectual basis for governmental restrictions on competition in such markets. On the other hand, if the market expands to a point where costs are no longer subadditive, or if technological change eliminates subadditivity, then average cost is lower when competitors are permitted to enter the market, and consumers are better off if regulation is diminished and competition is allowed to flourish.

Economies of scale refers to the behavior of cost as the production of a single product increases. Most modern firms, especially telecommunications firms, produce a number of products or services simultaneously, commonly using the same production facilities in order to reduce costs. If multiple products can be produced at less total cost by a single firm rather than by multiple single-product firms, all using the same technology, then *economies of scope* are said to exist. Let the multiple product total cost function be $C(Q_a,Q_b)$ where Q_a is the output of product a and Q_b is the output of product b. If $C(Q_a,Q_b)$ is less than $C(Q_a,0)$ + $C(0,Q_b)$, then a single firm can produce the specified amounts of the two products at lower cost than two firms each producing one of the products. The implication of the existence of economies of scope is that there may be a case for denying entry to competitors, thereby preserving the position of a multiproduct monopolist, even though the potential competitor can produce at least one of the products at lower cost than can or does the monopolist.

The question of subadditive cost in multiproduct monopolies and the existence of economies of scope are important in understanding whether a natural monopoly is sustainable in the long run in the face of competitive entry into certain of the monopoly's markets. A single-product natural monopoly dominates its market, and its price cannot be undercut by a competitor employing the same technology. In the case of a multiproduct monopoly in which one or more regulated prices exceed cost, possibly in order to subsidize other products, the monopoly is not sustainable against entry by competitors. Faulhaber (1975) has treated sustainability in detail.

Essentially, when a natural multiproduct monopoly exists, the sustainability argument suggests that regulatory policy should ensure that the monopoly is not subject to competition in markets where regulated prices exceed cost. For example, the partial deregulation of telecommunications markets in the United States in the 1980s created a situation in which regulated interstate toll charges were considerably higher than cost, and thus the regulated multiproduct industry was not sustainable against competitors or large customers utilizing bypass services.

The research on the sustainability argument for prohibiting entry into one or more of a monopoly's businesses was supported by Bell Laboratories, Faulhaber being a Bell employee at that time, and Bell sponsored much similar work over the years until the breakup in 1984 made such work irrelevant to its business goals. The argument rests on the twin assumptions that telecommunications technology is unchanging, and the monopolist, the Bell System in this case, is efficiently managed. The reader will have realized by this time that both assumptions are wrong.

It is important to distinguish the case where economies of scope may indeed justify denying entry to an efficient competitor from the case in which the potential competitor plans to sell a new product not offered by the existing monopolist. This important distinction was explicitly made by the FCC in approving MCI's application to offer a new product, microwave service, between St. Louis and Chicago over AT&T's objection that its monopoly would be nonsustainable.

Another argument used to support an existing monopoly position is that "common ownership over local exchanges, intercity facilities and manufacturing facilities is necessary for the efficient provision of telephone services" (Brock and Evans 1983). According to this argument only a single, integrated management can possess and properly use all of the necessary information required for optimal decisions throughout the system. Planning and control should encompass the entire network in order to achieve efficient, reliable and safe operations.

Evans and Grossman in Evans (1983) study the integrated management argument in an exhaustive fashion, concluding that, in the case of AT&T, management decisions were, in fact, decentralized and that integrated management offered no advantages over the competitive price mechanism when coordinating a nationwide telecommunications network. Market prices convey all the information necessary for efficient decentralized planning, and arm's-length economic transactions and contracts are sufficient for the coordination and operation of a nationally interconnected communications system.

National markets similar to telecommunications have operated efficiently for decades with individual business components under private ownership and control. As examples, consider the electric power industry, consumer electronics and electrical products, the air transportation industry, the automobile and supporting industries and the railroads. Coordination and standardization in these industries occur spontaneously in response to economic incentives rather than as the result of decrees issued by central management or government.

Rather than constitute a necessary component of an efficient national telecommunications industry, the integrated management of the Bell System probably slowed decision making and the innovation process while adding layers of expensive bureaucracy. As described in an earlier chapter, the breakup of the Bell System was promptly followed by a great deal of cost cutting and corporate downsizing in all of the new corporate entities.

Even more damaging to the integrated management argument is the development of new switched digital communications, in which a message can pass over many interconnected systems much like a freight car traveling over many railroads on its trip across country. Private PBX systems operating today under separate ownership and control provide flawless connections through regional and national telephone systems to other distant private PBX systems.

Another of the arguments favoring the monopoly market structure cites the advantages due to captive suppliers. It is claimed that the captive Western Electric Company responded faster and at lower cost to the equipment needs of the Bell System companies. This argument is patent nonsense since it is a violation of the "no free lunch" principle. The pricing of intermediate products sold among divisions of a corporation follows the transfer pricing principle, which states that the corporation as a whole maximizes profit when the transfer price is equal to the market price. There is no reason to believe that Bell operating companies gained any advantages by being restricted to making equipment purchases from Western Electric. The same sort of competitive arm's-length bidding by qualified suppliers that exists in most other industries would certainly satisfy the Bell operating companies' equipment needs just as promptly and reliably at the same or lower cost.

Rather, as noted by the FCC on different occasions, due to the regulated status of the Bell operating companies and the unregulated position of Western Electric, the latter's prices could easily and profitably be adjusted above competitive levels and the resulting higher costs passed on to consumers through the rate-setting process. The Department of Justice long sought the divestiture of Western Electric from AT&T on these grounds, and it is not possible to believe that consumers benefited from the exclusive purchasing arrangement.

AT&T often argued that a nationwide telecommunications system required a high level of equipment standardization in order to maintain safety and reliability. Such standardization required that any device connected to the system equipment must be manufactured by a Bell System subsidiary. This argument has been laid to rest by events. It is now clear that, just as in the case of the wide variety of consumer electronics products that must all be compatible with home and office electric outlets while meeting reliability and safety standards, the required level of standardization in telecommunications products is easily and efficiently achieved without government intervention by industry associations acting within the decentralized market system.

An argument based on financial benefits was sometimes made in favor of the former telecommunications monopoly. The claim was that the subsidiary companies of the Bell System enjoyed a lower cost of capital, presumably due to the financial strength and reputation of the parent organization. Such a lower cost of capital would benefit customers through lower rates. Keeping in mind the great size and stable revenues of the Bell subsidiaries, it is not likely that their capital costs would be lower as a result of common ownership. Modern financial theory does not support this claim, and careful studies have not shown any significant effects of such ownership on the cost of capital to the various companies (Brock and Evans 1983).

The Department of Defense supported the AT&T monopoly, sometimes on the basis of the supposed security advantages of its integrated paired-wire technology, at other times because, as a very large customer, Defense was an early beneficiary of the economical WATS service. It had been supposed that radio transmissions were more susceptible to interception than were hard-wire transmissions, but recently published histories of intelligence activities reveal that both technologies were and are vulnerable (Wright 1988).

As one considers and then rejects each of the arguments in favor of a national telecommunications monopoly, one wonders why the monopoly structure lasted so long in the United States.

The process of gaining complete control of the market may be accelerated and market control rigidly enforced if the monopolist becomes a regulated public utility. Once a monopoly becomes regulated, it acquires a powerful political constituency and a supporting dynamic often independent of the merits of the monopoly position. The regulators become wedded to the concept of a vertically integrated telecommunications monopoly, taking pride in their ability to understand the most intricate aspects of the regulated firm's operations and becoming the firm's advocates. Their positions as regulators, bringing them into frequent contact with industry leaders, lead to first-name relationships proudly revealed. Listen to Bernard Strassburg, a career FCC regulatory official: "No CEO was a stauncher defender and guardian of the Bell System's mission, purposes, structure and role in U.S. society than Charlie Brown. He had not embarked on his tenure as AT&T chairman to preside over the Bell System's dissolution . . . he remained steadfast in defending its vertically integrated structure and the part it was playing in the telecommunications revolution" (Henck and Strassburg 1988).

The old world of regulatory cooperation with business was civilized, stable and secure, totally unlike the new chaotic world of competitive markets and uncertain status. Regulation is predictable and has the force of law, while unrestrained competition generates surprises and may subject the public to unethical or illegal behavior on the part of sellers.

As described in earlier chapters, AT&T's strategy of monopolizing the national telecommunications markets was at first based on the takeover of smaller competing firms with justification provided by the theory of natural monopoly, and later based on its position as a regulated common carrier sustainable only if competitors were denied entry, and finally based on its long semipublic status as a national resource serving technological, defense and social objectives.

As a regulated monopoly, AT&T was not an active innovator and often used the natural monopoly argument to deny entry to competitors seeking to introduce innovations. AT&T, prior to the divestiture in 1984, frequently argued in hearings before the FCC (e.g., Brock and Evans 1983, 86) that new entrants into its businesses would duplicate expensive existing facilities and that such duplication is costly and not in the public interest.

This traditional natural monopoly argument is based on static economic reasoning, and is false when the potential new entrant offers to serve a market currently ignored by the monopolist or when the potential entrant intends to use a new technology permitting the public to be served at a cost per unit of service as low as or lower than the monopolist's cost, or when the new entrant, with its smaller management, is simply more efficient. In other words, the natural monopolist's argument that competition results in needless duplication of facilities is based on the assumption of unchanging technology and is only likely to be correct in the absence of technological progress, innovation and market growth.

Therefore, one may make the case that innovative activity undertaken by the regulated monopolist actually undercuts the argument for retaining its protected legal status, and therefore innovation is not only not actively pursued by the monopolist, but is something to be avoided or delayed. The large-scale acquisition of telecommunications patents for defensive purposes by Bell Laboratories over the years supports this argument, as do delays in the introduction of technologies such as FM radio. Only when entry is permitted does this incentive to delay innovation disappear.

When innovative products and processes are available to a potential competitor and there is a likelihood that such a competitor can serve the public at least as efficiently as the original monopolist can while expanding the market, then there is no economic justification for denying entry to the new firm. In fact, competition is likely to spur innovative activity on the part of the monopolist.

The FCC recognized the beneficial effects of such competition when it approved MCI's application to provide microwave service between Chicago and St. Louis in competition with AT&T. The FCC commented that the new microwave service would satisfy the unmet demands of "potential users who have no need for and cannot afford the services provided by AT&T" (Brock and Evans 1983). The commission noted that "the applicants are seeking primarily to develop new services and markets, as well as to tap latent but undeveloped markets for existing services, so that the effect of new entry may well be to expand the size of the total communications market."

In the U.S. long distance market, it is possible that the former Bell System had a strong economic interest in deterring new technologies such as switched digital data communications systems because of the large existing undepreciated investment in conventional facilities whose cost structure supported its claim to be a natural monopoly. As noted in Chapter 4, AT&T wrote off $6.7 billion in obsolete transmission equipment soon after the industry was reorganized and AT&T lost its monopoly status, resulting in the only earnings loss in 104 years. Counterarguments might point to the rapid expansion of microwave transmission in the AT&T network, although much of this capacity served specialized markets where competition was a significant threat.

The economies of scale of the newer telecommunications technologies are available at smaller service levels relative to the size of the market and, combined with the growth of markets, are much less likely to justify a monopolistic market structure. In the terms defined above, the new technologies are likely to change the cost structure from globally subadditive to locally subadditive.

As an example of one of the many instances of the effect of competition on innovation by a regulated telecommunications monopolist, when Datran proposed a switched digital data communications system, AT&T petitioned the FCC to deny Datran's application on the usual basis that approval would result in "uneconomic duplication of common carrier facilities" (Brock and Evans 1983). AT&T also argued that there would not be sufficient demand for Datran's services and that AT&T could meet the forseeable demand at lower cost. In response to Datran's initiative, AT&T accelerated development of its own digital data system, implicitly recognizing the economic merit of the Datran proposal. Datran, perhaps ahead of its time, went bankrupt, suing AT&T over alleged unfair business practices, but its proposed technology lives on in AT&T and in other systems.

CHAPTER 6 _____

The Economics of Telecommunications

INTRO

When significant economies of scale exist or to avoid needless duplication of expensive production and distribution facilities, an appropriate government agency will often authorize a single regulated public utility to serve the entire market in a geographical region so that consumers may benefit from the resulting low production costs. Classic examples of such regulated monopolies in the United States include electric, gas, water and telephone utilities, and various types of transportation and broadcast communication systems. Such monopolies are protected from competition by law, but the monopolist must also agree to government regulation of price and service.

In such cases it is important to determine the extent of economies of scale in order to ensure that the benefits of the natural monopoly structure are not outweighed by the costs and induced inefficiencies of regulation. It is also necessary to examine the proposed regulatory remedies to ensure that they efficiently serve the public interest rather than just protect the monopoly from competition.

Considering the effects of regulatory bureaucratization and the tendency of the law to emphasize equity over economic efficiency, many observers believe that such regulated monopolies have little or no incentive to introduce cost-cutting process innovations or product improvements. Government ownership of utilities is not much better with respect to the promotion of innovative activity. In late 1990 the government of Great Britain was reported to be opening its national telecommunications system, already privatized, to competition from firms in other countries, presumably to encourage innovation through increased competition.

The early telephone industry in the United States appeared noncompetitive because it did not seem to make economic sense to route more than one set of wires to each telephone customer, or to have more than one exchange in each local calling area. In the age of paired-wire telephone lines, all customers in a geographical area were best served by one local telephone company, and local companies preferred connection to one national long distance network. Not only did system reliability appear to require just one centrally coordinated national telephone network, but such a system also avoided needless duplication, thereby lowering costs to the benefit of the public. Since not all subscribers use the telephone system at the same time, the application of probability theory indicates that substantial cost savings may be achieved through reductions in capital equipment used in common. During the period of national expansion of the telephone network, the benefits of standardization appeared much greater than the potential benefits of permitting competitors to introduce competing innovative solutions.

Most other countries were served by government-owned telecommunications monopolies that were inefficient and unreliable in comparison with the Bell System. The U.S. model, in which one privately owned national company was regulated by federal and state governments, was generally recognized to be superior in performance to all other telecommunications systems around the world. Government agencies, particularly the Department of Defense, also saw benefits in a single national "hard-wire" long distance system with a long tradition of providing reliable, secure communications, particularly during the long period of cold war with the Soviets. Radio systems were thought more susceptible to interception and interference.

A large firm, as a natural monopoly, has a cost advantage over smaller firms and can engage in predatory pricing or force mergers, eventually gaining control of the entire market. Faulhaber (1975) concludes that actual predatory pricing, pricing below the cost of service, by a large firm seeking to monopolize a market does not make economic sense, and so true predatory pricing by monopolistic firms may be rare. More recently, Einhorn (1987) has shown that, in theory, it is optimal for a price-discriminating monopolist to price service below marginal cost for some large users as long as revenues in every customer class cover the costs of providing service to that class. However, regulated monopolists do not seem to be guided by marginal-cost reasoning, although there is no question that a large national firm encountering competition by a startup firm in some regional market has the financial resources promptly to lower its price and increase its advertising in that market. In theory, the competitive threat posed by the new firm can be countered by increasing its costs of doing business through competitive activities and litigation, thus eliminating the economic incentive to enter or remain in the industry.

Interestingly, when AT&T first contemplated the threat posed by the initial entry of MCI into a specialized microwave market, AT&T fought entry in the courts but decided not to retaliate by cutting price in that market because such action would spoil its pricing strategy in other markets in which similar services were provided. This supports Faulhaber's belief that predatory pricing by a monopolist facing competition in one of its markets is not likely.

The early history of the telecommunications industry reveals that a dominant firm controlling long distance communications can use its exclusive ability to provide such services as a means of gaining control of competitors in local service markets through merger. Recent regulatory changes indicate full awareness of this possibility, and when competition is permitted in some part of the telecommunications system, all competitors must have equal access to equal-quality facilities across the system.

Often the monopolistic market structure appears to make good sense to everyone, as in the case of an urban mass transit system, but sometimes a sophisticated analysis of technological options will reveal superior alternatives attainable only under a competitive market structure. An entrenched monopolist is in a position not only to raise price well above cost (where cost is defined to include a normal return or profit) and benefit excessively at the expense of the public, but also to deny the public the benefits of technological advances that might have been commercialized by competitors in a free market.

This chapter discusses whether or not economic data and other arguments justify the existence of a regulated telecommunications monopoly on a national scale, or justify the present system of regulated regional telephone monopolies. The discussion then turns to the economic effects of regulation in the telecommunications industry.

Since the existence of a natural monopoly is often considered the primary justification for government regulation, a great deal of interest has centered on whether the telephone industry in the United States actually exhibited the characteristics of a natural monopoly over its history, and whether the industry, or major parts of it, are natural monopolies today.

An industry may be considered a natural monopoly when the market for its goods or services may be served most efficiently by one firm (Meyer et al. 1980). Assuming the existence of a natural monopoly, a subsidiary issue involves the appropriate regulatory policy of government with respect to the industry, since regulation should have as its goal the provision of the maximum net benefits to society.

The net benefits of regulation depend on the magnitude of the economies of scale, the level of output of the firm when scale economies disappear relative to the size of the market, the market power, measured by consumer demand characteristics, of the monopolist and the likely effectiveness and costs of regulation compared to other remedies.

Regulation is supposed to eliminate redundant costs. If economies of scale are slight, the advantages of permitting some redundancy of service through competition, including the incentive for prompt incorporation of new technologies and the absence of regulatory costs, may outweigh the slight increase in the cost per unit of product due to the smaller sizes or redundant operations of the firms in the industry.

If substantial economies of scale exist but terminate at low levels of output relative to the size of the market, there is room for a number of firms, each operating at a cost-efficient rate of production, and effective competition can exist with little or no cost penalty to the public.

If there exist close substitutes for the product, the demand for the product becomes more elastic, or price sensitive, and the power of the seller to raise price above cost is decreased, lessening the need for price regulation.

As an example of alternatives to regulation, permitting free entry of new firms can reduce market power, and encouraging new technological developments can reduce the cost advantage of an existing monopolistic firm (Meyer et al. 1980).

A number of econometric studies have sought to gain evidence concerning economies of scale in telecommunications. Such statistical studies seek to identify the fundamental relationships generating the observed historical data. In particular, economies of scale are said to exist if the data reveal that equiproportional increases in all inputs to a production process yield a greater than proportional increase in output, holding technology constant.

Most of the studies cited here use time-series data for the period 1946–1975 and make simplifying assumptions about the technology employed, controlling for its change over time in order to isolate the effects of price or production rate changes. "Studies where comparisons are possible suggest that about one-third of the economies of scale that would be estimated if the impacts of technological change were not taken into account could in fact be due to technological change" (Meyer et al. 1980, 126). In other words, the economies of scale found in statistical studies may be overestimated by as much as one third.

A separate issue involves the usual assumption that technological change has an equal impact on both capital and labor inputs. This assumption may be faulty in the case of the telecommunications industry, and so the form of the production function used in the cited econometric studies may be incorrect (Meyer et al. 1980, 133).

The studies cited herein use aggregate historical data for the entire Bell System, but economies of scale are not expected to be present to the same extent in all components of a national telecommunications system, such economies being more pronounced in the long distance toll markets and less evident in local telephone operations.

Important data on variables such as physical output are not always available, and the investigator may have to rely on a proxy variable, such as value added, in place of some measure of physical output, in which case price changes can introduce subtle errors. Similarly, capital stock is used as a proxy for capital services, which is the correct variable, but for which data may not be available.

The statistical procedure used may fit a relationship to the "center" of the cluster of data points, in effect considering residual data variations to be random noise whose average value is zero. Alternatively, the statistical procedure may fit a relationship along the frontier of the cluster of data points thus differentiating between observations reflecting efficient and inefficient production. The latter approach is likely to be superior because of production inefficiencies introduced into the Bell System by regulatory procedures; for example, the tendency to tolerate obsolete capital equipment described earlier, but the former procedure is more common and more likely to be used.

Econometric studies of the telecommunications industry intended to determine the presence or absence of economies of scale usually estimate and report scale elasticities, defined as the ratio of the percent change in output resulting from a given percent change in all inputs. A scale elasticity equal to unity indicates constant returns to scale, implying that the cost per unit of service does not decrease as the system expands in size, and therefore also implying that there is no public economic benefit to be gained by restricting the market to a single supplier. On the other hand, scale elasticities significantly greater than unity indicate substantial economies of scale and tend to support the natural monopoly argument.

Three well-known econometric studies of the U.S. Bell system made in the mid-1970s estimated a range of scale elasticities from 0.91 to 1.25, indicating constant or slight returns to scale. Similar results were obtained in a separate study of Bell Canada, which reported scale elasticities in the range 0.85 to 1.11. (Meyer et al. 1980, 129). However, the FCC, after thorough review of the available econometric studies, did not conclude the the case for economies of scale in the Bell System had been proven, thus justifying its newly emerging position favoring competition in telecommunications. Sankar (1973) reviews a number of statistical studies of telecommunications cost data and, in his summary, reports constant returns to scale in the national Bell System.

The lengthy list of problems to be overcome in performing an econometric study of the telecommunications industry casts some doubt on the validity of the statistical studies cited, and these studies have therefore generated as much controversy as knowledge. Nevertheless, the cited studies do not suggest the presence of economically significant economies of scale in the national telecommunications system during the periods studied.

Instead of statistical data, engineering or simulation models may be used to shed light on the existence of economies of scale in telecommunications. Instead of estimating the coefficients and exponents in a production relationship using historical data, the engineering approach incorporates one or more mathematical equations representing or modeling the various components of the telecommunications system and their interactions as seen by an engineer designing such a system. Such equations may be manipulated in order to simulate the behavior of the actual system under study, or perhaps used to generate "pseudodata" for use in place of missing data in an accompanying statistical study.

Engineering studies have centered on the issue of economies of scale in long distance transmission since it is in this area that the greatest potential for competition was thought to exist. The basic cost components of long distance transmission are the costs of land, land rights and structures; the costs of "outside" equipment such as cables, antennae, and microwave towers; the costs of radio equipment, repeaters, etc.; the costs of multiplexers; and the costs of switching equipment. Investment cost studies often exclude switching costs and also occasionally exclude multiplexing costs, these exclusions affecting the results of the studies.

These cost studies are often performed at the request of sponsors seeking to generate support for a position in legal or regulatory matters. Multiplexing costs, for example, being approximately proportional to the number of circuits in use on a given link, can be a substantial part of total transmission costs, but there are little or no economies of scale in multiplexing. Therefore, the exclusion of multiplexing from a cost study tends arbitrarily to improve the case for economies of scale in long distance transmission, hence improving the case for a single national long distance telecommunications monopoly. Similarly, scale economies in switching equipment are minor and switching equipment is also often omitted from long distance cost studies, thus again arbitrarily increasing estimated scale elasticities.

Turning to the results of important transmission system cost studies: "Different types of of terrestrial transmission systems display an overall downward trend in investment cost per circuit mile as scale increases" (Meyer et al. 1980, 136). Including future technological options in transmission, the scale elasticity appears to be in the range 1.0 to 2.5, the latter value indicating very substantial economies of scale. Once again, these results exclude multiplexing and switching costs, which, if included, would tend to lower the range of scale elasticity values.

When multiplexing costs were included, one study determined that the scale elasticity was about 1.4, and this engineering simulation is consistent with actual AT&T cost data. "In sum, the engineering evidence suggests that economies of scale in long distance may not be too much greater than the range found in econometric studies of the whole telephone system. The long distance scale elasticity appears to be between 1.1 and 1.5 at the outside" (Meyer et al. 1980, 140).

Evans and Heckman (Brock and Evans 1983, 141) argue that such engineering studies give a false picture of the existence of economies of scale because they do not include managerial and other costs that, due to bureaucratic inefficiency, are likely to increase with system size. They suggest that "these studies ignore the costs of managing the firm and thereby ignore the possibility that managerial diseconomies may outweigh engineering economies." They note that engineering economies of scale exist in many industries that are otherwise not considered natural monopolies. While AT&T management argued that integrated system planning can reduce costs over the whole network, actual planning in AT&T was decentralized. The system planning economies that may be available in theory do not seem to be taken advantage of in practice.

It is important when conducting a study of productivity to separate the effects of technological change from scale economies. Several companies can realize savings from technological change as easily as can one company. The separation of technological change from scale effects is difficult to accomplish, however, and so estimates of scale elasticities are probably biased upwards.

Satellite communications were not included in the above cost studies. Small-scale satellite communication systems are attractive now, and satellite circuit costs are expected to decline even more in the future as technology advances. But total satellite capacity is small relative to terrestrial capacity, and so scale economies in satellite communications add little to the scale economies existing over the national telecommunications system.

There appear to be very little scale economies in customer-premises equipment (CPE), which is consistent with the findings for consumer goods industries such as stereo equipment and small home appliances.

In summary, the existence of economies of scale in the former national telecommunications system significant enough to justify its monopoly structure has not been established.

The discussion in this chapter has thus far emphasized the economics of the production of telecommunications services. It is also necessary to examine demand characteristics and pricing policies in telecommunications, comparing the effects of such policies on the consumption of telecommunications services in order to assess the effectiveness of regulation.

The voluntary exchange principle underlying the theory of competitive markets is that all buyers and sellers should be perfectly free to respond to prices as they wish, producing or consuming whatever products and services they prefer. The aggregate actions of consumers, each maximizing the satisfaction derived from consumption, and the aggregate actions of sellers, each maximizing profits, produces an equilibrium market solution in which aggregate consumer demand equals aggregate supply at some market price.

No coercion is involved when markets are competitive since any consumer may refuse to buy at the market price and any producer may refuse to sell at the market price. Government regulation inevitably brings coercion of some sort into the operation of the market, either by imposing a regulated price above or below the market price or by constraining production or consumption decisions. Since the regulatory policy may actually be a politically expedient response to external pressure groups, which are not as concerned about the performance of the market in question as they are about their own objectives or economic status, it is often an accident when regulation improves market performance. Regulatory policy is usually developed and applied with the constant advice and assistance of self-serving attorneys, consultants, accountants and other expert witnesses and intervenors representing the regulated firm or pressure groups.

The nature of the services offered by a modern telecommunications company must be considered to gain an understanding of the effects of price and other variables on the demands for these services.

The telecommunications services commonly provided are:

1. the installation and operation of equipment providing a subscriber the option to access the telephone network on demand;

2. the use of the system while in communication with some other party;

3. the option to access any other subscriber including, for example, health and public safety organizations;

4. the provision of a unique telephone number enabling the subscriber to be called by all other subscribers;

5. new computerized products such as data services, recreational opportunities and information services; and,

6. services of social value such as civil defense, education and participation in government

The telephone usage service described in category 2 may be further classified as residential (leisure) usage and business usage. These usage classes, in turn, may be divided into local calls and long distance (toll) calls.

The prices charged for these services by telecommunications firms should emulate competitive market prices, because competitive prices, through the process of voluntary exchange, provide the most benefit to society. In practice regulated prices in the former Bell System diverged substantially from cost-based market prices for many reasons.

Competitors were quickly attracted to the specialized long distance toll market when MCI was permitted to enter, indicating the regulated long distance tolls were well above the prices that would prevail in a competitive market. In contrast, no competitors wished to provide local residential telephone services, which were cross-subsidized by revenues from long distance operations and thus provided below cost. In the last few years, the rapid rise in local charges to replace the reductions in subsidies from long distance as that market became more competitive have created incentives for competitors to enter the local telephone market, especially using new technologies.

The amount of each of the above services used by an individual depends on the prices charged for the services, the income of the individual, the prices of complementary or substitute communication services, and the tastes of the individual, which are likely to be heavily influenced by employment, age and family status (Wenders 1987).

Telephone calls have different values to different callers and call recipients; the business call quickly closing an important deal before conditions change or the emergency call to a hospital can be extremely valuable, while the casual social or business call may have little monetary value and may even be a nuisance to one of the parties.

For any given subscriber and type of call, there will generally be a tendency to respond to an increase in the price of telephone usage by making fewer, or shorter, calls. Similarly, higher-income individuals or members of families with higher incomes tend to make greater use of the telephone. Not only do higher-income subscribers have more discretionary funds, but the time spent in communicating, and thus saved by telephoning rather than communicating by other means, is more valuable to them. Telephone use also tends to be a factor in achieving and maintaining a high-income status. Thus there is a complex of interdependent reasons explaining higher telephone use by high-income subscribers. Finally, studies have shown that the time spent using the telephone, everything else being equal, is higher for teenagers and lower for subscribers over 55.

Empirical studies of the consumption of telephone services confirm these general principles and also provide quantitative estimates of the sensitivity of such behavior to economic data such as price and income. Empirical studies commonly report estimates of various elasticities, the percentage changes in consumption that can be expected per unit percent change in one of several relevant economic variables.

The price elasticty of demand for residential access has been reported to be quite small for the United States as a whole, about -0.039 (Perl 1983). This means that each 1% decrease in the residential charge for access, the fixed part of the monthly telephone bill, tends to increase the number of residential subscribers by 0.039%, a very small response. This low-elasticity estimate is not at all surprising and indicates that the former regulatory policy of encouraging universal residential telephone access through subsidies was probably not warranted. In other words, the decision to subscribe to residential telephone service is not significantly affected by the price of such service, at least in the price ranges commonly encountered in the United States. Supporting this 1983 empirical study, residential access prices have generally increased since the breakup of the Bell System in 1984, due largely to the nationwide loss of residential access subsidies, with little or no apparent effect on the number of residential telephone subscribers. The price elasticity of demand for access appears to increase as income decreases, as might be expected. Therefore, lower income tends to be associated with a greater sensititivity of demand to price, but demand is still inelastic, falling in the range -0.2 to -0.3.

The income elasticity of demand for access is estimated at about +0.5, just about what might be expected. For example, a 10% increase in income is likely to result in a 5% increase in the demand for telephone access. The combination of very little price elasticity and moderate income elasticity for access to telephone service indicates both the importance of telephone service and its lack of substitutes, and therefore explains consumer resistance to rate increases (Wenders 1987).

The demand for single-line access by small businesses is probably even less price elastic than is the demand for residential access.

Turning to the second type of service, local residential usage appears to have a low price elasticity in the range -0.27 to -0.38. Since local residential usage is subject to many sociological factors including family size and background, the effect of income on usage is not clear (Brandon 1981). Low-income families may be more sensitive to the perceived requirements of social acceptability than are middle-income families, and may use a greater variety of telephone services as a result. High-income families are large consumers of telephone services.

Better data is available concerning the sensitivity of intrastate and interstate toll usage to income and price. Intrastate toll usage short-run price elasticities averaging -0.21 have been estimated, with long-run price elasticity estimates in the range -0.22 to -1.04. Intrastate income elasticities have averaged -0.39 in the short run and +1.33 in the long run. Therefore, intrastate usage is much more sensitive to price and income than is access.

Similarly, interstate usage price and income elasticities are substantial, the long-run price elasticity estimated to be about -1.0 and the long-run income elasticity probably exceeding unity. Consequently, interstate toll usage is even more sensitive to price and income than is intrastate usage, at least for residential customers. Price elasticities appear to increase with distance, the effect of a given price change to affect usage becoming greater with the distance between the parties to the call. Business long distance usage is less sensitive to price, surely price inelastic, and residential usage is more sensitive to price.

Since the emergence of competition in the long distance markets, flexible plans that coincide with these empirical results have been advertised by all of the major long distance companies. For example, in place of the old AT&T toll charges per minute of long distance service, varying with the time of day and distance between parties, the company now offers plans like the "Select Saver," a fixed monthly charge plus a discounted price reflecting usage per call.

The monopolistic seller of a good or service, by definition, possesses the ability unilaterally to raise the price charged, customers having no alternative but to pay or do without the service. The provision of an important product or service under monopolistic conditions has traditionally been seen as justification for some sort of government intervention; either direct government regulation or indirect regulation based on the threat of antitrust action by the Department of Justice, or even nationalization of the industry. Economists have come to believe that a careful study of the likely benefits and costs of government regulation must be made in order to determine which of the available government regulatory options, including the option of doing nothing, are in the public interest. For example, considering the significant costs of regulation, the option of doing nothing may be preferable if a noncompetitive market is contestable, in which case a dominant firm is likely to be effectively deterred from significant price manipulation by the threat of entry by new firms. Such entry is impossible if the monopolist is protected from competition by its regulated status.

In theory, the ideal form of government regulation would bring about industry behavior resembling that which occurs under competition, while providing positive net benefits to the public, considering the costs of regulation.

According to Wenders (1987):

While competitive behavior is usually sought under regulation, true competitive behavior is not well understood. The common impression is that, under competition, all firms in the industry are earning just enough profit to keep them operating at their present levels of activity, but not enough profit to encourage additional firms to enter the industry or to encourage new investment. This is a long run equilibrium solution in the case of a static, constant technology market. But high technology industries like telecommunications provide constantly recurring opportunities for the introduction of new technology into production processes and into the various service products, and the required investments will not be made unless above average profits are likely.

Above-average profits temporarily arising due to innovations are a natural outcome of an efficient, evolving competitive market. The existence of such profits should not be considered a reason for regulating an industry. When regulation is implemented, the regulatory system must permit firms to capture such short-term profits as incentives for innovation.

As Wenders (1987, 203) puts it,

When prices are above the [market] long run supply curve, abnormally high profits will be earned by new capital and thus investment will be encouraged. In addition, the lure of potentially high profits will induce suppliers to try new products and means of production. Thus, the possibility of above-cost pricing provides a lure for innovation, and at the same time the fact of above-cost pricing encourages the competitive investment that will ultimately prevent it from lasting.

However, it has been extremely difficult to devise a politically acceptable regulatory scheme that simultaneously protects the public from price gouging while providing economic incentives that encourage innovative activity leading to product improvements and cost reductions by the regulated firm. In 1988 the FCC implemented a "price cap" regulatory strategy intended to resolve this problem in the case of AT&T, but it is too soon to tell if this new approach will be effective.

During the period in which the national expansion of telephone service occurred, the coordination and standardization provided by integrated management was of great public benefit. The relatively low volume of service in the regional and interstate markets, the simple types of service sought by customers, and the economics of the original telephone technology also supported the concept that telephone service should be provided by a regulated monopoly, in control of all aspects of service from end to end.

However, with much less reason, the Bell System argued that it should also have a monopoly over telephone equipment manufacturing and own all customer-premises equipment. The telephone equipment business monopolized by Western Electric, being unregulated, offered opportunities for abuses of market power, and this part of the business was always a source of conflict between AT&T and the Department of Justice, playing a prominent role in the final breakup of the Bell System.

The FCC did not share the Department of Justice's concern over the vertically integrated structure of the Bell System. The FCC supported AT&T when the company fought the attachment of "foreign" equipment, no matter how innocuous, to Bell System telephones:

The Hush-A-Phone case is a classic illustration of the regulatory values that dominated the entire telephone regulatory community for generations. They were embraced by the FCC from its beginnings in 1934 and into the late 1960s. Thus, it was the conviction of the FCC and its staff that they shared with the telephone company a common responsibility for efficient and economic public telephone service, and that this responsibility could only be discharged by the carrier's control of all facilities that made up the network. . . . It also had to extend to any equipment a subscriber might attach to or interface with the basic service (Henck and Strassburg 1988, 38).

The Communications Act of 1934, establishing the FCC, was clearly intended by Congress to provide checks on the economic power of the Bell System in particular while maintaining order and encouraging progress in the broadcasting industry in general. While agreeing in the past with the natural monopoly concept of telephone communications services, the federal government had long sought the divestiture of Western Electric from AT&T on the grounds that the combined companies enjoyed excessive and unwarranted economic power and that telephone equipment could be provided to the public at lower cost under competition. An indication of the difficulties facing government regulators and Justice Department officials is that it took almost 100 years and enormous expenditures of time and money on litigation by government and private parties to separate Western Electric from the Bell System. Even after the historic breakup of the Bell System in 1984, Western Electric is still a subsidiary of AT&T. Western Electric was certainly not a money loser for AT&T.

There is little controversy over the principle of government regulation of monopolistic industries in the United States today, but there is intense controversy over the definition of monopoly, the extent and cost of government involvement in markets and the philosophy that should guide such involvement. Perhaps most important, when government intervention in an industry occurs, the form of such intervention is not determined by the technical analysis of some supposedly omniscient expert, but is determined in the political arena, where concepts of economic efficiency are often muddied by political realities, the body of law and the often complex facts of each case.

Professional regulators working in the real world of political trade-offs are often amused by the theoretical regulatory prescriptions of academic economists. The latter, in turn, are continually frustrated by the divergence between government regulation in practice and government regulation in theory, and often document the extent and social cost of the divergence in their scholarly writing.

A natural assumption is that government regulation is instituted to improve the general welfare. There are a number of theories that attempt to explain the regulatory process, some quite startling in their candid appraisals of the importance of parochial political interests and bureaucratic inertia in government regulation of industry. Serious discussions over deregulation may sometimes center on, for example, what will happen to the jobs of the regulators. Some studies assume that regulators are interested only in the public welfare, but astonishingly, there exists little evidence other than occasional anecdotes to support this view.

Less information is available on the effects of various forms of government regulation on the rate of technological change and innovation, a primary interest of the present work. With few exceptions, economists, as social scientists, often either simplify the process of technological change or assume that it is somehow absent in the course of using static models to study other dimensions of market behavior.

Innovation has dramatically reduced the cost of providing telecommunications services and brought to market many valuable new products. Government regulatory policy has changed accordingly, recognizing the importance of encouraging industrial innovation. Telecommunications play a vital role in a modern economy centered on the use of high technology in the delivery of services. However, it is not only the rate of innovation in telecommunications services and products that is important but also how innovation in telecommunications affects the rate of innovation and hence productivity and competitiveness in the economy as a whole, a question addressed in a subsequent chapter.

The apparent breakdown in the free market system in the 1930s provided public support for a greatly enlarged role for government in the business sector. The belief that government could and would solve the persistent problems of society was prevalent, and political astuteness of governmental agencies, combined with bureaucratic inertia, often lead to the continued existence of many governmental functions long after it was recognized that they were ineffective.

By the 1980s, a consensus had emerged in the United States to the effect that government regulation often imposes excessive costs on industry, including the insidious hidden costs of technological stagnation. Such hidden costs were revealed by the dramatic growth of business competition across national boundaries, exposing inefficiencies and technological stagnation in regulated and unregulated industries alike. The costs of government regulation, in terms of induced inefficient behavior, often appeared to outweigh regulatory benefits and, as a consequence, the political climate began to favor deregulation. Important industries were deregulated, and the process seems to be continuing, although in a selective way.

It is now commonly thought that it is better to avoid government regulation even if the industry in question does not quite satisfy all of the requirements of competition. It is in the context of a national mood favoring deregulation of industry combined with a worldwide revolution in electronics that the breakup of AT&T and the transition of the telecommunications industry to some mixed form of workable competition and regulation must be seen.

The monopolistic firm, under rate-of-return regulation, the prevailing form in the United States, is permitted to earn a fair return on the assets in its rate base. The regulated firm thus has an incentive to depreciate its capital equipment slowly and retain it in use, perhaps alongside new equipment installed to service market growth, long beyond the time it would have been retired by a competitive firm. There is no generally accepted way to ensure either that economically obsolete capital equipment is excluded from the rate base or that total undepreciated capital investment is not excessive. Many regulated public utilities operating in the United States today continue to use economically obsolete capital equipment alongside modern equipment, earning the same return on both, the higher operating costs being passed along to customers through higher rates.

As technological progress occurs over time, potentially more efficient competitors intending to employ new low-cost technologies are inevitably attracted to the monopolized business by the prospect of high profits. A monopolistic firm must be able to keep such competitors out of its market. As previously described, the Bell System, through its philosophy of acquisition of patents, cooperation with regulators, and active defense of its regulated status in the courts, used government itself as the primary means of deterring potentially more-efficient competitors and maintaining its monopoly. For example, radio and microwave technologies, not requiring expensive land lines, were a low-cost competitive threat to the conventional telephone system with its enormous investment in cables and associated equipment. The Bell System was never a pioneer in the innovative uses of these new technologies, and its legal warfare against MCI is a matter of history. The average level of telephone technology in service in the United States is in danger of falling behind that of other countries (Hanley 1989). At least one new urban telephone system in Eastern Europe is being based on cellular radio communications rather than on conventional paired-wire technology. The Japanese are reported to be planning a $200 billion national fiber-optic telecommunications system.

The form of regulation common in the United States and the desire to be considered natural monopolies provide incentives for regional telephone companies to continue using existing paired-wire analog technology for local telephone service even as superior alternatives come into use elsewhere in the world. Therefore, it is important that regulation address economically efficient depreciation policies and alternatives to the rate of return on assets method of calculating rate adjustments.

The Bell System built a national telephone network with a reliability envied all over the world and brought modern, economical telephone service to most markets in the United States. Nevertheless, in recent years certain classes of customers were in fact being charged rates well above cost, and all customers were increasingly denied the advantages of technological progress and efficient operations. By the 1970s times were changing, and the telecommunications system that had served the nation quite well for years was falling behind in terms of product development and cost of service. Technological progress was rendering obsolete the argument that telephone service is a natural monopoly. Voluminous court records, statements by AT&T officials and recent actions of the company, now facing the hard realities of a competitive market, all support this conclusion.

The economic argument against permitting competition when economies of scale exist is that the monopolist may experience slower growth and not move as quickly down the declining average cost curve. Therefore, permitting competition would seem to slow down the decrease in the cost per unit of service and thus cost consumers more in terms of higher rates. This argument against competition assumes:

1. The monopolist is operating efficiently at present so that the pressures of competition will not lower the monopolist's cost per unit of service.

2. Those costs of the regulatory apparatus borne by consumers are insignificant.

3. The rate of technological progress is independent of industry structure so that any technological advances that might be introduced by the new competitors would also be introduced by the monopolist.

Based on the history of the telecommunications industry summarized in earlier chapters, it is this writer's opinion, probably shared by the great majority of observers, that all of these assumptions are false. In any case, the economies of scale in telecommunications that may have existed in the past are eroding due to technological advances (Meyer et al. 1980, 148), and the market power of the regional telephone monopolies is also decreasing as substitutes for conventional telephone communication are developed.

Government regulation is intended to bring about an improved market structure, but there is doubt that regulation generally achieves this objective. There are many instances, for examples, trucking and airlines, where regulation has failed to improve market behavior. In local telephone service, where regulation of entry is intended to avoid the high costs of redundant installations of competing conventional copper-wire loop networks, cellular systems and satellite systems can avoid such redundancy while providing substitute services that reduce the market power of the local telephone monopoly.

Another spurious argument in favor of regulation is that it encourages the monopolist to provide service to out-of-the-way locations that would not otherwise be served, and to do so by subsidizing rural rates with profits earned on low-cost, high-density service, all customers paying essentially the same rates. The social goal of universal telephone service can supposedly be best served by a monopolist charging rates based on system average cost and subsidizing the high-cost rural markets. Competitors, presumably choosing to enter only profitable high-density markets, would "skim off the cream" of the business and leave the monopolist unable to provide service to rural customers at reasonable rates. However, rate uniformity in other than interstate long distance rates may be a mirage rather than a reality, and AT&T is not heavily involved in serving truly low-density rural locations today and never was (Meyer et al. 1980, 173).

In summary, the arguments for restricting entry and regulating the telephone industry, or the regional telephone companies, as natural monopolies are getting weaker with the passage of time and advancing technology. The agreement to break up the Bell System did not include any explicit reduction in the regulatory powers and practices of the FCC or the state regulatory commissions. The political climate in recent years has encouraged competition as well as movements toward deregulation which are expected to continue in the future.

CHAPTER 7 _____

The Theory and Practice of Government Regulation of Telecommunications

Although the Bell System was dismantled in 1984, seven regional holding companies coming into existence in addition to the downsized AT&T core company, there was no accompanying dismantling of regulatory policies and procedures at either the state or federal levels. Regulation of the telecommunication industry will continue to evolve in coming years, and the transition from regulation towards competition will play a significant role in the growth of markets and the rate of technological advances in telecommunications. A background for studying the effects of regulation on the rate of innovation in telecommunications can be established by a brief review of the historical and legal bases for utility regulation in the United States and a description of current regulatory practice. Several modern theories of utility regulation are then offered.

When substantial economies of scale exist in an industry and it becomes clear that the market is most efficiently served by a single firm, the question of the proper role of government in the regulation of the monopolistic firm becomes important. The first issue to be decided is whether government will take control of the firm's assets and operate it as a government enterprise, or whether the firm will continue as a private entity with its policies and pricing decisions subject to approval by a regulatory commission at the state or federal level. The latter model, being consistent with the status of private property under the Constitution, is more prevalent in the United States, especially when the regulated firm is economically viable and not likely to require government subsidies.

Joseph Chamberlain, writing in 1894 about the problem of utilities in Great Britain, as quoted in Crew and Kleindorfer (1986, 95), commented:

The supply of gas and water, electric lighting and the establishment of tramways must be confined to very few contractors. . . . As it is difficult to reconcile the rights and interests of the public with the claims of an individual, or a company seeking as its natural and legitimate object the largest attainable private gain, it is most desirable that . . . the municipality should control the supply, in order that the general interest of the whole population may be the only object pursued.

Government ownership and operation of the national telecommunications system, often as a part of the postal system, became the most common model in Europe. In the 1980s the Thatcher government in Britain liberalized telecommunications by privatizing British Telecommunications PLC, resulting in improvements in the quality of service and lower costs in that country.

Government regulation of private business in the United States has a firm legal basis. In *Munn* v. *Illinois* the Supreme Court in 1877 first established the principle that government has the right to regulate a private enterprise when the public has an interest in the use to which the property of the enterprise is put. In a later case Justice Bradley stated:

We held that when an employment or business becomes such a matter of such public interest and importance as to create a common charge or burden upon the citizens; . . . when it becomes a practical monopoly, to which the citizen is compelled to resort and by means of which a tribute can be exacted from the community, it is subject to regulation by the legislative power. (Crew and Kleindorfer 1986, 97)

In a case involving the pricing of milk, *Nebbia* v. *New York*, 94 US 113 (1877), the Supreme Court ruled that an industry is subject to regulation even if it is not monopolistic: "A State is free to adopt whatever economic policy may reasonably be deemed to promote the public welfare."

As Crew and Kleindorfer (1986) note, "Implicit in these legal decisions has been the notion that . . . the monopolist's price should reflect average cost, both to protect the consumer as well as to ensure that the firm did not incur a loss in its operations." Average cost means the firm's total cost divided by the quantity of product or service sold. Total costs are the sum of all operating expenses plus the sum of all capital costs, the latter including principal and interest payments on debt, preferred stock dividends and common stock dividends. This common regulatory policy, in which the regulated price is set equal to average cost, is intended to yield a fair rate of return on the firm's assets and is therefore also called "rate-of-return regulation." From the standpoint of economic theory, the guiding principle of voluntary exchange implies that the regulated firm should not be forced to provide service at a loss, and average-cost pricing is intended to achieve this objective as well as the objective of fair consumer prices.

Economists have long quarreled with the concept of rate-of-return regulation in regulated industries because, while fair to the firm, it does not encourage economically efficient production or consumption decisions. Perhaps the most famous article on this topic was written by Averch and Johnson (1962), who showed that rate-of-return regulation creates economic incentives for the firm to use resources inefficiently, particularly capital resources. But economic efficiency is not necessarily the primary goal of regulation. Zajac (1978) concludes that the public and trained economists have different concepts of the objectives of regulation. The public, to whom regulators feel responsible, believe that equity, fairness and social justice are the proper objectives of regulation.

Historically, economic justice has been the driving force behind regulation. According to one well-known writer (Rawls 1971), the principles of justice require that each person has an equal right to the most extensive scheme of equal basic liberties compatible with a similar scheme of liberties for all, and social and economic inequalities must be to the greatest benefit of the least advantaged, and offices and positions must be open to all under conditions of equality of opportunity.

Glaeser (1957, 196), writing on the origins of government regulation, states that "all attempts at such regulation owed much to a very ancient ideal of social justice, which, as applied to economic life by the early church fathers, became their very famous doctrine of justum pretium or just price." According to Glaeser, the primary concern of government regulation, probably influenced by the prominent role of the legal system in the United States, is fairness and equity rather than economic efficiency. Zajac (1982) affirms the tendency of the regulatory process to emphasize economic justice at the expense of efficiency, and goes further, claiming that regulation emphasizes the status quo, a principle of justice that pervades the legal system. Thus, regulators can be expected to turn down reforms that while benefiting the public in general may harm the status of some segment of the public.

Theodore Vail, the early general manager of AT&T who set the course followed by the company for 100 years, also favored rate-of-return regulation. Vail, writing in AT&T's annual report of 1907, in an era when European nations were forming government-owned telephone companies, recognized the necessity of some form of government regulation: "It is contended that if there is to be no competition there should be public control. It is not believed that there is any serious objection to that control . . . provided that capital is entitled to its fair rate of return . . . and enterprise its just reward."

The "fair rate of return" concept was confirmed in the Hope case. Justice Douglas concluded: "From the investor or company point of view it is important that there be enough revenue not only for operating expenses but also for the capital costs of the business . . . the return to the equity owner should be commensurate with the returns on investments in other enterprises having corresponding risks" (Crew and Kleindorfer 1986, 98).

Vail, in AT&T's annual report of 1910 stated: "It is believed that the telephone system should be universal, interdependent and interconnecting, affording opportunity for any subscriber in any exchange to communicate with any other subscriber of another exchange . . . It is not believed that this can be accomplished by separately controlled or distinct systems nor that there can be competition in the accepted sense of competition."

The concept of regulation proposed by Vail, the regulated company accepting no more than a fair rate of return on its investment in return for being granted an exclusive franchise, became the form of regulation that governed AT&T and most other public utilities in the United States over the years.

Thus, regulation is intended to protect the consumer from the threat of monopoly and also to protect the rights of the utility and its owners, but these objectives are often achieved more in form than in substance. The potential for rent seeking is great under rate-of-return regulation. Rent seeking consists of those activities undertaken and resources expended in order to alter the distribution of economic gains arising from the artificial scarcity created by government regulation. Therefore, rent-seeking activities not only add nothing to economic welfare but reduce it by needlessly consuming resources. Rent-seeking activities constitute a large part of the legal maneuvering that accompanies the filing of a rate case with a utility commission. The regulatory process not only contains no built-in incentives tending to limit such socially valueless expenditures, but probably encourages rent seeking.

By providing a fair return on capital, rate-of-return regulation has eased the problem of raising funds for investment in capital-intensive utilities: "While regulation does not seem very concerned with promoting efficiency, either in a traditional static sense or in the sense of providing incentives for research and development, it does achieve a stable basis for operations which have . . . resulted in companies being prepared to make the investments required"(Crew and Kleindorfer 1986, 101).

Our attention now shifts from the legal and historical basis for utility regulation to a review of several of the most important current theories of regulation. A theory of regulation is necessary in order to understand the regulatory process, to be able to predict its behavior under various circumstances and to be able to evaluate the likely effects of different forms of regulation on technological change and innovation in telecommunications.

There are many theories of government regulation of public utilities, but the present discussion considers only samples from three groups: the public-interest theory, the capture theories and a group of miscellaneous theories.

The public-interest theory of regulation is based on the idea that regulation should maximize the public interest, defined in economic terms as the sum of the monetary values of the net benefits provided to society by the regulated firm. The net benefit provided to each customer is the difference between the value the customer places on the service, or reservation price, and the cost of providing the service to the customer, including a competitive return to the owners of the firm's capital. The theory is normative (or prescriptive) rather than positive (or descriptive) because it specifies what regulatory behavior should be rather than providing an explanation that fits the facts of existing regulatory behavior. The public-interest theory therefore assumes that regulators require utilities to make pricing and resource-allocation decisions that maximize the economic dimension of social welfare, defined as the sum of net consumer and producer benefits.

Since competitive markets provide incentives for participants to make decisions that, in the aggregate, tend to maximize social welfare, the public-interest theory of regulation essentially assumes that regulators promote economically efficient decision making in regulated firms emulating the behavior of competitive industries. More succinctly, monopoly is viewed as a "market failure," an aberration from competitive behavior, which can and should be altered so as to return the affected industry to the competitive norm, or as Crew and Kleindorfer (1986, 105) put it: "Using regulation to perform the function of competition by inducing the monopolist to act like a perfect competitor."

The fundamental principle of the public-interest theory of regulation is that price should be set equal to marginal cost, the theoretical guiding force in competitive markets. When the demand for service varies cyclically over time, peak and off-peak times can be identified. Since marginal cost includes the opportunity cost of fixed capital investment and this opportunity cost is zero during the off-peak period and reflects fixed costs during the peak period in an optimally designed system, marginal cost should vary according to time of demand. This economically efficient approach is called peak-load pricing, and there is substantial literature on this topic. STOP

There has always existed extensive controversy over the precise definition of marginal cost in various contexts, and the legal background of regulation favors average-cost pricing, as described above. Consequently the "marginal cost pricing" rule, almost universally favored by economists, may be complex in application and was rarely encountered in the regulated U.S. telecommunications industry.

It is difficult to sustain a theory of regulatory behavior when the theory prescribes a form of behavior that is hardly ever observed. An equally serious but less obvious problem with the public-interest theory is that it assumes that regulation is costless. Any improvements in utility operations brought about through regulation are assumed to be net benefits to society. In fact, the costs of regulation often outweigh the benefits achieved through regulation (Posner 1975, 819), so that in many cases society may be better off not attempting to regulate a monopolistic firm. The fact that regulation still exists in those cases tends to undercut the theory that such regulation is in the public interest. Recent modifications of public-interest models include treatment of the costs of regulation, for example, an interesting model offered by Crew and Kleindorfer (1986).

Most economists would agree that the prescriptions offered by the public-interest theory of regulation, centering on the importance of marginal cost in pricing and resource allocation decisions, are correct, in spite of that theory's lack of applications as a descriptive theory.

The general absence of the type of regulatory behavior predicted by the public-interest theory resulted in the development of various "capture" or "private-interest" theories. The "Economic Theory of Regulation," for example, views the outcome of regulation as a product or commodity desired by various competing private groups; for example, consumers, employees, shareholders, managers and perhaps even the regulators themselves in the sense that their budgets depend on their production of "regulation." The benefits of regulation are awarded to those who are able to convince the regulators, in one way or another, that they value it the most. It is usually assumed that the regulatory preferences of a small number of individuals, each with a great deal at stake, will prevail over the preferences of a large number of individuals, each with little at stake. Consequently, the "capture" type of theory tends to support the idea that regulation favors large firms at the expense of consumers, although it is not unusual to find regulation favoring a special consumer group.

Wenders (1987) notes that the capture theory is not consistent with the history of regulation of the telecommunications market. The long-standing national policy of cross-subsidization of local exchange non-traffic-sensitive costs with revenues based on interstate toll charges runs counter to the theory. The interstate toll customers are mostly large companies, and the capture theory predicts that these companies would cause regulation to develop in a manner favorable to their interests. Except for the eventual appearance of some high-volume discount services in the interstate market, this never occurred.

However, in this case all of the state regulatory commissions, many of whose members face regular elections, are subservient to consumers as voters. Consumers are reminded of the cost of telephone service each month when the bill arrives. The state regulatory commissions, which are members of the National Association of Regulatory Utility Commissioners (NARUC), present a politically potent and united front to the FCC, which has generally responded by approving regulation in the interstate markets favoring residential consumers over business. Congressional action, by and large, has also supported the interests of residential consumers over that of business, most recently in delaying the implementation of the FCC's access charge plan.

Peltzman (1976) describes the demand and supply sides of the market for regulation, the demand side often consisting of a group of companies desiring regulation as a superior form of cartel management, provided by regulators and imposed with the force of law. Industries and professions often lobby to be regulated in the interests of market stability and control. For example, as described earlier, Theodore Vail willingly accepted government regulation on behalf of AT&T at the beginning of the century, and from that point on AT&T monopolized what had been a competitive telephone industry and also monopolized the telephone equipment manufacturing and distribution industries.

However, one might argue to the contrary that Vail sought regulation not because it was preferable to the existing market structure but as an alternative to government ownership, the trend in Europe at that time.

One can also cite numerous examples of regulated firms not being permitted to earn the allowed rate of return or having certain costs and investments disallowed. Finally, the capture theory assumes that regulation takes place behind closed doors, whereas in fact important regulatory issues are often decided in an open judicial setting with qualified counsel representing consumer interests.

The political science version of the capture theory emphasizes the role of pressure groups in obtaining control over the regulatory process. There is certainly a great deal to be said for the realistic nature of the political science model, but this version is not believed to be as operational, that is, as useful for predictive purposes and as efficient with respect to information requirements, as is the economist's version.

The third class of miscellaneous regulatory theories includes, as examples, the behavior to be expected from regulatory bureaucracies (Niskanen 1971), the requirement for a contract between utility and customers, supervised by the regulatory agency, prior to the utility undertaking the necessary large investments (Goldberg 1976), and a means for reducing risks and preserving the status quo in the face of technological change in utility markets (Owen and Braeutigam 1978).

As mentioned, average-cost pricing is a very common regulatory philosophy in the United States in spite of the fact that, from a public-interest standpoint, the theoretically ideal form of regulation is marginal-cost pricing. Marginal-cost pricing may be difficult to use for at least three reasons:

1. Declining average and marginal costs in a natural monopoly, with marginal cost necessarily lying below average cost at the quantity of product demanded, means that marginal-cost pricing would not recover sufficient revenues to allow the firm to earn a fair rate of return.

2. Marginal costs are more difficult, that is, more costly, to measure, varying greatly depending on the time of day and the definition of the product and terms of service, and therefore such pricing may be impractical.

3. Prices based on marginal cost may appear unusual or unfair or run counter to social objectives such as providing subsidies to some classes of customers.

It is interesting that regulatory apportionment of total costs among customer classes to achieve social objectives is also consistent with the monopolist's tendency to practice price discrimination in order to maximize the sum of the revenues obtained from all customer classes. The concept that customers who value the service most, as indicated by their low price elasticity of demand, should pay the highest rates has a long theoretical history and is often favorably considered by regulators when trying to resolve difficult rate cases while raising adequate revenues. Prices that depart from marginal cost in inverse proportion to price elasticity so as to meet revenue requirements are called Ramsey (1927) prices.

The multipart tariff, in which the price consists of a flat monthly charge plus a charge per unit of service consumed, is another common method used to set a price that reflects the cost of providing a unit of service while ensuring that total revenues cover the total costs of the regulated firm. Essentially, the flat monthly charge should reflect the non-traffic-sensitive costs of providing access, while the usage charge, again reflecting cost, applies to each call made. It is interesting to observe that simple flat monthly charges for residential service, with little or no charge for usage, was the most common model when the industry was completely regulated, but local exchange companies are moving towards the multipart tariff as local markets become more competitive.

Average-cost pricing is inefficient in the sense that such prices do not encourage socially optimal consumption decisions. Average-cost pricing is also inefficient in the production or "X-efficiency" (Leibenstein 1966) sense because it provides no incentive for the firm to use resources in a socially optimal fashion. Average-cost pricing provides the same incentives and can be expected to achieve the same results as the "cost-plus" pricing philosophy once common in defense contracting. In the context of the early telephone industry, the average-cost pricing objective meant that rural customers paid, in theory at least, the same rate as did urban customers, even though the cost of providing the service to the rural customer was much greater. Similarly, high-volume business users did not receive a discount for many types of service even though the cost of providing the services in volume was much less. The inherent subsidies granted to certain classes of subscribers under average-cost pricing were justified to state and federal regulators on the grounds of encouraging universal telephone subscription and service.

When several identifiable classes of customer exist, for example, residential, commercial and industrial, total capital and operating costs are apportioned to each customer class on the basis of cost causation. The total cost incurred to serve each customer class is then divided by unit sales to that class to arrive at an average cost for each class upon which the rate for the class is based.

There is usually some scope for discretion in the apportioning of costs among customer classes, because much of the firm's capital equipment is used in common to provide service simultaneously to all customer classes. Therefore, in seeking sources of revenue to meet the firm's costs, the firm's apportionment of its common costs, approved by regulators, may result in average cost, and hence the regulated price, being highest for that customer class with the lowest price elasticity of demand.

The regulatory process begins a new cycle when the utility files an application for a rate increase. The decision to file a request for a rate increase is usually based on the utility's inability to earn the previously allowed return on common equity, given existing rates and other operating and capital costs. The company must present accounting and other data as evidence to justify the need for a rate increase, such evidence usually documenting the unsatisfactory return being provided to shareholders under existing rates.

Public utility commissions use historical cost as well as accounting information on sales and operating expenses to determine the carrying charge for debt and preferred stock. The cost of equity, however, must be estimated for each rate case. In estimating the cost of common equity the commission will consider testimony from expert witnesses on what investors expect to earn on the utility's common equity. Interested parties, called intervenors, and commission staff are also permitted to file testimony on the expected return on equity. Outside expert witnesses, present at the request of the regulated utility, usually provide estimates on the high side and the commission staff and consumer advocate, if one exists, provide estimates on the low side. The commission then decides on an allowed return on common equity that, when combined with all other categories of costs, and when the separations and settlements process, described later, is completed, is used to establish new rates for the different classes of service. In some cases an agreement satisfactory to all is not reached, and the rate case is decided in court.

The separations, settlements and division-of-revenue process refers to the policy of allocating a share of the property and expenses of local telephone service to the interstate toll markets. Advances in long distance voice transmission technology over the years have been accompanied by significant cost decreases. On the other hand, the costs of providing local service have not decreased. The national policies of universal telephone service and rate averaging, as well as political pressures, set limits on the growth of local exchange rates. The revenue shortfalls of the local exchange companies were increasingly met by a complex process of cross-subsidization from interstate toll revenues. Since all of the business entities were subsidiaries of the Bell System, such cross-subsidies were handled by internal accounting procedures, and these internal transactions did not affect the overall financial position of AT&T.

The separations and settlements process is so arcane, has such a specialized jargon and has such a convoluted history that discussions of the question in print are often almost impossible to comprehend. Yet, it is important to describe this process because of the fundamental role it has played in the development of the telecommunications industry in the United States and because it remains the source of serious problems facing the industry.

The principle behind separations and settlements rests on the fact that local telephone facilities are used when making long distance (toll) calls, and so a portion of the long distance revenues should be remitted back to the local companies. Since local telephone facilities are used in common for both local and toll calls, the costs of such facilities are common to both types of service. Except for a special case (Weil 1968), there is no theoretically correct way to allocate common costs, and so a complex of accounting procedures have been developed over time to solve the separations problem. The revenues collected from the separations process are turned over to local telephone companies in the form of settlements, which often amount to a significant fraction of local company revenues.

Separations refers to the policy of separating both the traffic- and non-traffic-sensitive portions of local telephone company accounting costs into a part that remains in the state jurisdiction in which an exchange is located, and a part that is assigned to the interstate market under the jurisdiction of the FCC, where these costs are added to the costs of providing interstate toll services for purposes of setting interstate toll charges.

The nationwide telecommunications investment base subject to separations was about $164 billion in the middle 1980s. By law, telephone companies are entitled to set rates that provide sufficient revenues to cover expenses and also to ensure a fair rate of return on this investment base. Telephone companies perform jurisdictional separations to conform to the nation's regulatory structure. The FCC has regulatory authority over all interstate communications, and state governments regulate intrastate services through public utility commissions. Each government jurisdiction regulates different services, and rates for those different services are influenced by overall revenue requirements. However, while jurisdictionally distinct, the different telecommunications services are provided to the local subscriber by means of the same plant and equipment. A local access line can be used to make both interstate and intrastate calls. Under the separations and settlements process the investment in the line must be separated between the two jurisdictions for regulatory and ratemaking purposes.

Historically, the separation of plant and other expenses between interstate and intrastate services in order to establish rates began with a dispute involving railroads. The Minnesota rate cases, decided by the U.S. Supreme Court in 1913, established the basic principle that property costs should be allocated to each jurisdiction according to the relative use made of the property. For example, the subscriber line use (SLU) definition of the fraction of local non-traffic-sensitive costs to be assigned to the interstate market is simply interstate toll minutes divided by the total of all toll and local minutes of use.

Prior to 1930 the separations process was guided by the board-to-board theory. The cost of long distance service was considered an add-on to local service. All local exchange costs were associated with the provision of local service and were covered by local revenues. Interstate costs consisted of only those expenses associated with operating and maintaining the interstate facilities connecting local exchanges.

In 1930 the Supreme Court in *Smith* v. *Illinois Bell Telephone Company* ruled on telephone separations for the first time. The Court decided that the separation of interstate and intrastate property, revenues and expenses is essential to the appropriate recognition of the competent government authority in each field of regulation. Up to that time separations based on the board-to-board theory did not include the non-traffic-sensitive costs of connecting subscribers to the telephone network, such costs constituting the majority of local telephone company costs.

The first Separations Manual, the separations guidelines contained in Part 67 of the FCC rules, was published in 1947. The Separations Manual applied the SLU definition given above to local exchange non-traffic-sensitive costs, thus increasing the share of fixed local costs borne by the interstate jurisdiction. Between 1947 and 1971 periodic revisions to the Separations Manual had the effect of further increasing the fraction of local exchange costs assigned to the interstate jurisdiction.

The Ozark Plan of 1971 introduced the subscriber plant factor (SPF) for the separation of non-traffic-sensitive subscriber plant. The SPF factor was approximately 3.3 times the SLU factor and, once again, greatly increased the share of local costs allocated to the interstate jurisdiction. The 1971 Ozark Plan classified plant as either traffic-sensitive (TS) or non-traffic-sensitive (NTS). TS. costs vary with the usage of the facilities and are therefore variable costs. NTS costs do not vary with usage and are therefore fixed costs. For example, if 10% of a local company's total traffic is interstate, then approximately 33% of the company's NTS plant would be assigned to the interstate jurisdiction under SPF. As a result, high-volume toll users are assessed many times the cost of access through toll rates, thus providing a strong incentive for them to bypass the common carriers. The fixed costs of the local exchange have been subsidized by toll charges on long distance usage, and the demand for long distance service over the common carrier networks is inversely related to the magnitude of such toll charges. Therefore, the separations and settlements process created very large economic incentives for high-volume users to leave the interstate common carriers at the same time that advancing technology, combined with rational pricing, should have been making the interstate network more and more attractive to such users.

Increasing numbers of high-volume users have been, in fact, employing the latest technologies and imaginative business arrangements to bypass the local exchange companies, with potentially severe consequences for the low-volume and occasional interstate toll users, whose rates must necessarily increase.

In recognition of this problem, the SPF was frozen in 1981. In 1982 the FCC, in Docket 78–72, attempted to reduce the growing difference between long distance toll charges and costs, but Congress, apparently concerned about adverse political reactions to the consequent increases in local telephone rates, prevented the FCC's plan from going into effect (Wenders 1987, 165).

"The initial FCC access charge plan was designed to promote competition, prevent bypass and ensure the continued availability of affordable local-exchange service" (Danielsen and Kamerschen 1986, 80). Access charges, as they exist thus far, are fixed periodic charges on end users intended to cover at least part of the costs incurred by local exchange companies in providing access to interstate transmission facilities. Revenues from access charges permit reductions in the payments from interstate carriers to local exchange companies and thus permit reductions in interstate toll charges, bringing such charges closer to competitive levels.

Since access charges fall on end users and are therefore politically sensitive, a compromise access charge plan was instituted in the early 1980s, and access charges are still in a transitional state as this is written. While interstate toll charges remain above competitive levels, uneconomic entry is encouraged in the interstate market, and the interstate common carriers have incentives to respond to this competition in uneconomic ways, including, but not limited to, bypassing the local exchange companies.

An important associated problem with access charges is that, while such charges are determined by regulators, they are an increasingly important component of the competitive strategies of the local exchange companies, who must be free to adjust these charges in accordance with market realities. Access charges, like all prices, should reflect the market value of the services provided. If necessary to generate sufficient revenues, access charges should depart from cost to a greater extent in less-elastic markets. End users setting a high value on having access to interstate carriers should pay a higher charge than do occasional residential end users. Access charges should properly reflect volume of use, time of use and perhaps the directions in which calls are made (Danielsen and Kamerschen 1986, 80). It is important that state and federal regulators and the public realize the necessity for pricing flexibility as competition plays a larger and larger role in the telecommunications market.

Interstate subsidies to local exchange companies grew out of the desire to ensure universal telephone service in a monopolistic system, but these same subsidies in a competitive environment are now causing interstate business to bypass the local exchange companies, thus threatening local exchange revenues and possibly bringing about the local rate increases that will endanger universal telephone service.

It was the practice of charging toll rates substantially higher than cost and remitting part of the resulting excess revenues to local companies, along with the rather inflexible attitude of AT&T towards the needs of large customers, that initially attracted competition in some toll markets. Large corporate users confronted with what they considered excessive costs simply built their own telecommunications networks and bypassed AT&T. Competitors such as MCI built long distance transmission systems based on new technologies but relied on local telephone companies to provide facilities for originating and completing calls.

Now that AT&T has competition in the interstate toll market, its competitors, the Other Common Carriers (OCCs), are also required to remit portions of their revenues to local telephone companies. However, the FCC, in order to encourage competition in the toll markets, set the competitors' charges at 45% of those paid by AT&T. These discounts provided to the OCCs are scheduled to decrease over time and finally disappear.

The FCC favored a policy of nationwide rate averaging with differentiation for volume, distance and time of use, but since costs vary geographically, rate averaging has resulted in underpricing some routes and overpricing others. The high-density–long distance routes were probably the most overpriced, and it was these routes that attracted competition.

The states have also practiced cross-subsidization by pricing state toll services and customer-premises equipment above cost in order to subsidize some classes of local service. This practice is called residual pricing. Because statewide rate averaging is commonly practiced, the extent of the state subsidy to local service varies widely, probably being highest in rural, low-density areas. Since many studies and common sense show that the decision to have a telephone installed in the home is not price sensitive, such subsidies have been a politically popular windfall to a very large group of voters.

As indicated, subsidies often flow from high-density urban customers to low-density rural customers. A common justification of this policy is value-of-service pricing. The value-of-service principle is based on the idea that telephone service is valued differently by different classes of customers. A customer valuing telephone service highly should pay a higher price than a customer who sets a lower value on telephone service. Therefore, according to this principle, urban customers, who are presumed to value telephone service highly, ought to pay a larger portion of local exchange costs. Value-of-service pricing also comes into play when pricing business and residential service. Here, the business customer is presumed to set a higher value on telephone service and is charged accordingly, to the benefit of the residential customer.

There is no support for value-of-service pricing in economic theory, except for cases in which prices based on relevant cost do not generate sufficient revenues to provide a fair rate of return. The necessary departure of prices from cost should then follow the value-of-service principle in order to achieve the least distortion of the quantities of services demanded. Therefore, before instituting value-of-service pricing, there should be clear evidence that prices based on cost will not provide sufficient revenues to support the service.

A reason often offered for the subsidization of local telephone service is that such service generates positive externalities and therefore should be priced below cost. *Positive externalities* refers to the benefits flowing to other parties as a result of a particular individual or family having telephone service installed. In theory, the sum of these external benefits, not normally considered when an individual is deciding whether to have service installed, should be subtracted from cost when pricing service. Prices corrected for externalities provide economic incentives resulting in a socially optimal aggregate demand for telephone service. However, as Wenders (1987, 5) and many other writers comment, there is no evidence that actual subsidies to local telephone subscribers are based on measurements of these externalities.

The preceding discussion centered on the regulation of rates charged by telecommunications companies. Regulatory authorities control another important economic incentive—the depreciation of plant and facilities. Depreciation permits the initial investment cost of plant, facilities and equipment to be considered a series or schedule of annual expenses over future years. While depreciation schedules should have some basis in the declining economic value of the facilities and equipment over time, more often such schedules satisfy accounting, tax and regulatory objectives. Reserves accumulated through the capital recovery process often constitute a significant part of the funds needed for capital replacement in local exchange companies. Current regulated depreciation rates evolved over the years in which the telecommunications industry was a regulated monopoly and, for the most part, are inadequate in an era of competition.

In a regulated environment, a slow rate of depreciation is preferable from the company's standpoint since it economically enlarges the investment base on which the regulated rate of return is earned, and since prompt capital replacement is not a high priority due to the lack of competition. As markets become competitive, such slow depreciation schedules interfere with the ability of a company to respond to technological initiatives by competitors. In the new competitive environment the old reasons for slow depreciation are forgotten. A Southern Bell executive complains: "For instance, fully 37% of the useful life of Southern Bell's plant and equipment has been consumed, but only about 18% of the asset value of that plant and equipment is accounted for in its depreciation reserves." According to this executive, depreciation is believed to be a major stumbling block in the path of Southern Bell, hindering its ability to become fully competitive (Frank Skinner in Danielsen and Kamerschen 1986, 87).

Summarizing telecommunications policy in the United States under regulation, the overriding concern at the national level has been the separations and settlements process for the allocation of costs such that a large part of the local revenue requirements were assigned to the rate base of the interstate toll markets. At the state level the dominant features of rate making were residual pricing, statewide rate averaging and value-of-service pricing. Flat-rate charges for local service were common, and usage charges for local calls were unusual. Interstate and intrastate toll charges were greatly in excess of cost. These politically popular practices could be supported because the cost of providing toll calls was declining due to technological advances, resulting in increasing net revenues in the toll markets.

CHAPTER 8 _____

Modern Telecommunications Technologies

It is difficult to appreciate the tremendous technological change that has been occurring in telecommunications in recent years and the impact of this change on market structure without delving into the technology to some extent. In this chapter the telecommunications technologies employed in the recent past, the present, and those likely to be put into practice in the near future are described in layperson's terms. An understanding of the new and emerging technologies and their economic significance helps clarify the recent revolutionary developments in the telecommunications industry and highlights likely future trends.

The first communication system to use electricity to transmit messages was the telegraph, which still plays an important role in specialized, high-volume applications. The simplest telegraph system consists of a pair of stations, each with a telegraph key, connected by two wires. At the sending station, the operator momentarily depresses a key to connect a battery to the two wires, causing a pulse of current to flow. The momentary current flow in the wires is detected as it passes through an electromagnet at the receiving station. The electromagnet, a piece of soft iron surrounded by a coil of wire, rings a bell or sounds a tone or click. Communication occurs by using the key to send a pattern of short and long pulses of current to spell out letters and numerals. The code invented by Samuel Morse was in universal use for this purpose until the 1920s.

The classic telegraph system is very slow, prone to errors because different letters are represented by pulse patterns of different lengths, and requires human operators at each end. In later systems telegraph keys were replaced by teletypewriters and the Murray code was used. Each letter or numeral typed by the operator was automatically represented by a pattern of five electrical signals. Each signal consists of the presence or absence of a pulse of current. Since each Murray code letter or numeral is the same length and the code is generated and received by machines, errors are reduced. Teletypewriter operators achieve speeds of 60 to 100 words per minute, although the international motor standard is 66 words per minute. In the early teletype systems operators used teletypewriters to punch patterns of holes in a paper tape. The tape was fed into an autotransmitter which ran continuously in order to make maximum use of the telegraph line. At the receiving end a second paper tape was punched for retransmission of the message if required. Alternatively, a printed copy of the message, the "telex," was made from the punched tape using a reading machine. Modern electronic teletype systems use various codes and data transmission speeds.

Teletypewriters are frequently connected to automatic exchanges or switching centers such as the International Telex System. The number of the receiving teletypewriter may be dialed just as in a telephone system. The punched paper tape has since been replaced by computer equipment, and electronic switching is used in modern teletype systems. Because each message may be stored in electronic form and transmitted with other messages at an appropriate time, modern teletype systems make very efficient use of transmission links. Reliable, economical and efficient modern teletype systems continue to be popular for high-volume business and government communications. STOP

A telephone station consists of the familiar telephone receiver and its receptacle, one or the other usually containing a rotary dial or set of push buttons for dialing. The telephone handset is linked to the telephone exchange by a two-wire line that forms a current-carrying loop. Speech is transformed by the microphone in the telephone into current modulations, which are carried by the two wires to the receiving telephone where the current modulations are transformed back into speech. This type of sound-conversion technology is called an analog system because the current modulations vary smoothly and proportionately with the sounds of speech. An additional wire is usually present for control and signaling purposes. When the telephone is in its receptacle, the current that carries speech is cut off, and only a small alternating current is available to produce a ring when the station is called.

When the telephone receiver is picked up to make a call, current begins to flow through the speech loop, and this current is detected by equipment at the local exchange. The exchange responds by sending a signal that the caller hears as a dial tone. The caller then dials the number of the recipient of the call. Each digit dialed by the caller is transmitted as a series of clicks or pulses by the older handsets, and as pairs of tones in the newer telephones. A subscriber loop interface circuit (SLIC), located between the handset and the exchange, supplies power and the ringing current.

The first telephone systems were simply a pair of instruments connected by a wire loop. Exchanges or switchboards were quickly incorporated into telephone systems, permitting the caller to reach any other station connected to the exchange. The locations and the number of exchanges are determined so as to minimize the total length of the connecting subscriber loops while taking advantage of economies of scale in switching, thus minimizing system cost.

Originally, the human operator at the exchange physically connected the lines of the two parties, thus completing the speech-carrying circuit. Operator errors, slow response times and the high cost of making manual connections led to the invention of the automatic exchange.

A Chicago undertaker, Almon Strowger, unhappy with an unreliable telephone system that appeared to be costing him business, invented an automatic telephone switching device in 1889. Strowger's device is central to automatic electromechanical switching systems, which continued to be based on his ideas up until the advent of electronic switching systems in the late 1960s.

Suppose there are no more than 100 subscribers connected to the system. Then each subscriber requires only a two-digit telephone number. Strowger's device consisted of a stationary hollow cylinder with ten parallel rows of electrical contacts on its inside surface, each row containing ten contacts. The 100 contacts inside the cylinder are each connected to a wire passing through the wall of the cylinder. The wires are in turn connected to the one hundred telephone lines of the subscribers.

As the caller dials the first number, represented by a sequence of up to nine electrical pulses, a pickup arm moves up the same number of rows inside the cylinder. The second number dialed causes the pickup arm to move the same number of contacts along the selected row, the pickup arm stopping at the contact connected to the recipient's line. An additional wire, as mentioned above, provides signaling and control functions. Current can then flow through the speech loop to the recipient's telephone, which rings. The connection, a physical wire path for the voice-carrying electrical current, is maintained during the conversation for the exclusive use of the two parties. Until the conversation is complete and the telephones are replaced in their receptacles, "the line is busy."

When more than 100 telephones are connected to an exchange, the usual situation, multiple Strowger type group selectors are used in a set of switching equipment, one selector for each digit dialed.

Going into a little more detail, a modern automatic electromechanical switching system works in the following way. When the telephone receiver is picked up, a rotary electromechanical switch called a uniselector begins to move, looking for a free set of switching equipment among the several thousand that may be available at the local exchange. Each of the sets of switching equipment at the local exchange is based on the Strowger system just described.

The subscriber's telephone is connected to the first set of free switching equipment encountered. If none is free, a busy signal is heard. Supposing a free set is found, the first unit of the set of switching equipment connected to the subscriber's telephone is called a group selector, and it provides a dial tone to the subscriber and also energizes a control wire warning the uniselectors of other subscribers that this set of switching equipment is in use. The first group selector moves its two pickup arms to the two contacts indicated by the first digit of the telephone number. Two arms are needed for the two speech-carrying wires. The first group selector also finds and connects the subscriber's line to a second group selector that will deal with the second digit of the phone number dialed. This process is accomplished while the subscriber is still dialing the second digit. The second group selector makes the contacts required by the second digit of the telephone number and also finds a free group selector to handle the third digit of the number.

The process continues until the complete telephone number has been dialed and a connection has been made to the recipient's line, at which point a ringing generator in the exchange causes the recipient's telephone to ring. The ring is also heard by the caller over a separate wire because the recipient has not yet picked up the telephone.

When the recipient picks up the telephone, current flows in the two speech-carrying wires, the connection is complete, and the ringing tone generator is disconnected. As complicated as it sounds, everyone who has used a telephone knows that most of the time it works quickly and effortlessly. Up until the 1980s Strowger-based systems were used to switch over half the world's telephone lines because these systems are flexible, cheap, reliable and easy to comprehend. Strowger electromechanical systems are disappearing in favor of solid-state electronic switching.

The exchange, or automatic switchboard, connects the calling station loop to the loop of the recipient if both are connected to the same exchange. If the call is out of the local area, connections are made between one or more additional exchanges, the last exchange ringing the addressee's telephone. A physical wire path from caller to recipient is therefore arranged by automatically setting a sequence of appropriate switches in the network for the exclusive use of each call. The higher the volume of telephone calls passing through an exchange, the greater the number of connecting wire paths that are required, and, of course, the greater the cost. Also, a smaller number of longer calls can require as many wire paths as a larger number of short calls, and so the intensity of traffic depends on the product of the number of calls and the average length of each call. The design capacity of the telecommunications system, determined by the number of trunk lines and sets of switching equipment, is based on a statistical analysis of the projected traffic intensity.

System cost increases in a roughly proportional way with system capacity. But notice that capacity, when based on probability theory, need not increase as fast as the number of subscribers, since relatively few of the subscribers use the system at the same time. For example, an exchange with 10,000 subscribers might require only 2,000 sets of switching equipment, and the amount of switching equipment per subscriber tends to decrease in larger systems.

Thus, there appear to be economies of scale in local exchange operations, and this is usually the primary justification for considering local telephone companies to be natural monopolies. Two competing local telephone companies would require twice the amount of switching equipment and subscriber lines, and consumer rates would necessarily be much higher. It is not clear that such economies of scale in local telephone operations will continue to exist when newer digital switching and digital signal technologies are employed.

The further apart the parties to a call are, the greater the length of the required wire path, and so cost also increases with distance in traditional telecommunications systems. This is not necessarily so in the most modern systems using digital switching or using microwave or satellite technology. Using satellite technology, the signal goes up 22,300 miles and then returns 22,300 miles, so the distance between stations on earth is inconsequential.

When the caller begins to speak, the sound is converted into an analog signal that modulates the current in the wire loop. The current modulations are picked up by the listener's telephone and converted back into sound. Considering that the speed of tranmission of electrical signals is far, far faster than the production of sounds during speech, nothing much is happening most of the time when two parties are conversing over the wire path connected for their exclusive use. The automatic teletype system with message queuing described earlier makes much more efficient use of a telecommunications network than does a conventional telephone system, which is why they were popular for high-volume business communications.

The automatic electromechanical switching system has worked well in telephone networks throughout the world, but there are some drawbacks. The moving parts in electromechanical switching systems occasionally fail or reduce the quality of sound reproduction. Such switching systems require a clean, dry operating environment, plenty of floor space, constant maintenance, and they are relatively expensive.

Electronic exchanges, widely introduced in the 1970s, replace electromechanical switches with solid-state transistors and integrated circuits that have no moving parts, operate much faster, are very reliable, provide a higher quality of sound transmission, are very small and require less power.

Whereas the electromechanical exchange sets up a sequence of switches as the call is being dialed, the newer electronic exchanges first store the digits being dialed and ring the recipient's phone, waiting for it to be picked up before connecting a path through the network for the call. No path is tied up if there is no answer or the recipient's line is busy. Alternative routes through the network are automatically tried if necessary. Electronic switching enables long distance international calls to be connected much faster and completed at lower cost. A whole range of new services, for example, pushbutton dialing and answering features, can be provided using electronic exchanges, the telephone system becoming much more reliable and user-friendly.

The concept that each conversation requires the exclusive use of a physical wire path was satisfactory in Alexander Graham Bell's time, but there must be a better way. In fact, making use of modern electronics technology, there are a number of better ways.

Carrier systems increase the number of conversations that can be carried by a single pair of wires. Many high-frequency carrier waves, each modulated by the voice signals of an individual conversation, may be simultaneously transmitted over long distances using the conventional wire pair. In the United States, 12 to 24 telephone channels are usually carried on two pairs of wires, one pair used for speech going in each direction. Each conversation requires only 4kHz of bandspread, so speech-modulated carrier waves of 8 kHz, 12kHz, 16kHz and so forth may be used, each carrier wave transmitting an individual conversation. Telephone conversations are converted back into the normal 0 to 4kHz voice range at the receiving end of the long distance system.

Instead of pairs of copper wires, coaxial cables may be used in carrier systems, greatly increasing the number of voice channels. A coaxial cable consists of a wire in the center of a copper tube supported by insulating material, and this type of cable configuration increases the frequency response to cover a very wide part of the spectrum. A long distance coaxial telephone cable may contain 12 coaxial units surrounding a core of numerous conventional wire pairs. Depending on the specific configuration and band width, coaxial cables can carry over 1,000 analog telephone conversations simultaneously, but actual installations are designed for the expected volume of traffic, which may be much lower. For purposes of system reliability, long distance coaxial cables often parallel microwave relay systems. While the wire-pair cable is only suitable for voice and data transmission, coaxial cables, with their greater frequency response, can also carry video signals which contain about 250 times as much information as do voice signals.

Electronic switching is a great improvement over electromechanical switching, but both transmit speech by analog means and require that an exclusive circuit be connected between the caller and the recipient. The analog transmission of the conversation makes poor use of the telephone system, which is capable of transmitting information at far greater rates.

There are new ways of seeing and representing certain natural phenomena. Since before recorded history we have perceived and represented the world in analog terms—using pictures. The general theory of information, developed by Dr. Claude Shannon of Bell Laboratories, enables us to break down the analog entities of our normal perceptions (such as sight and sound) into separate digital bits. The digital bit language is intelligible to machines and can be quickly manipulated by computers with no loss or distortion. By 1982 the digital phonograph was a reality, and by the end of the eighties the CD (compact disc) had almost completely supplanted LP records. The price of CD players has declined from about $1,000 in 1983 to about $200 in 1990.

Digital technology for representing sound, a further improvement over electronic switching, has dramatically changed the way sounds are transmitted as well as the way telephone calls are switched and routed. In general, using digital switching technology, exclusive circuits forming transmission paths are not needed. Conversations can be electronically converted into digital signals, groups of nondenominational zeros and ones, which may be stored in buffers, that is, electronically compressed and periodically transmitted as groups or packets of data in digital form. Time-division multiplexing allows the transmission of bursts of digital data representing one conversation in between other groups of digital data carrying other conversations over a shared network. From an economic standpoint, the distance between caller and recipient is no longer the critical system dimension; it has been replaced by time. "The evolution toward entirely digital systems has virtually no disadvantages, particularly from the moment when the majority of the links are digital. Most of the intermediate equipment is eliminated. In addition, 'temporal' automatic switchboards are far less expensive than 'spatial' automatic switchboards" (Pujolle 1988, 8).

The central advantage of packet switching is that it is not necessary to arrange a unique wire path through the system for each conversation by closing a sequence of switches, either electromechanical or electronic, across the network. All of the circuits remain connected all the time, and microprocessors at junctures direct message packets along the most direct open routes. Segments of the conversation, converted into groups or packets of digital data, can each be given the correct address and sent to find their own easiest way through the network much as different groups of children can have their destinations pinned to their jackets and sent traveling across country. Some of the children going to the same destination may travel by different routes depending on conditions existing in the transportation network.

Because the speed of transmission of electrical signals is so much faster than our relatively slow speech, the packets of conversation in digital form all arrive in order at the destination, like infinitely tiny bowling balls returning to the rack after play, and are electronically converted back into the voice of the speaker. Not only is the process far faster and cheaper than the analog approach, because the same physical system is used more efficiently and can carry much more traffic, but the quality of sound reproduction is superior. Digital signals, each consisting of the presence or absence of a pulse of current, are much less sensitive to the quality of the transmission system than analog signals. Therefore, more communication channels may be accommodated over the same physical network, each channel being used far more efficiently and providing higher quality sound transmission.

Both computer data and voice transmissions may be sent over the same channel, and so a personal computer can remain connected to a remote data source by a telephone line while the same line is being used for telephone calls. Local networks providing such digital communication services to government agencies, large businesses and educational institutions were common in the United States by the end of the 1980s.

The widespread availability of inexpensive integrated circuits makes all this not only possible but economically practical. But physical wire paths, possibly consisting of coaxial cables time-shared by many simultaneous users, are still involved. Alternatively, high-volume communications may be transmitted over long distances using microwave radio relay stations or using satellite communication systems, the extremely short radio waves employed being suitable for carrying information-intensive video signals.

By the turn of the century specialized wireless communication systems were being developed. While wireless communication, or radio, has many limitations, the technology is particularly useful for communication when one or both of the stations are moving, as in the case of ships at sea, or, more recently, automobiles.

Guglielmo Marconi is usually credited with the invention of radio, having received the first patent on June 2, 1896, and, later, the Nobel Prize in physics for his work. However, Alexander Stepanovich Popoff, a lecturer in physics at the Russian Imperial Navy's Torpedo School at Kronstadt near St. Petersburg, was conducting successful radio experiments at about the same time. Popoff is considered to be the inventor of radio in the Soviet Union, but Marconi's work led directly to the rapid commercial development of wireless communications in the last years of the nineteenth century. While the name of the inventor is uncertain, many scientists being interested in radio at that time, it is clear that Marconi was the innovator. Marconi was most interested in marine communications between ships and shore stations. The availability of radio communications quickly proved valuable for acquiring assistance in the event of collisions and other disasters at sea and in naval operations in World War I.

Radio waves are propagated by causing alternating electric current to flow in an antenna, the higher the frequency of current oscillation the shorter the length of the radio wave. Low-frequency radios require long antennas, sometimes hundreds of feet in length, and a great deal of power. Such systems are bulky and not very portable. But low-frequency radio waves follow the curvature of the earth and may be used to communicate over great distances. In the 1920s amateur radio enthusiasts were permitted to use the high- frequency section of the radio spectrum because it was believed useless for long distance communications.

The amateurs discovered that inexpensive shortwave radios, with their low power requirements and short antennas, could be used to communicate across the Atlantic Ocean, the waves "bouncing" off, or being relayed by, the ionosphere layer in the upper atmosphere rather than following the curvature of the earth. Because the reflection of radio waves by the ionosphere is affected by solar activity and does not occur at all at very high radio frequencies, other techniques, such as networks of land-based relay stations, have been developed to provide reliable worldwide radio communications. The very high frequency and ultra high frequency parts of the radio spectrum have since been extensively exploited for a wide variety of purposes.

Radio frequencies at the upper end of the spectrum are so short that very small antennas may be employed. Multiple small antennas may be fed by a single transmitter and grouped so as to reinforce and focus the generated radio waves, forming a high power beam that may be directed to a particular receiving station.

High-frequency radio waves may be used for carrying telegraph, voice and data information but are not suitable for carrying video information. The transmission of live television pictures requires frequencies in the microwave spectrum, and these ultra small waves travel in straight lines right off the surface of the earth and out into space. Geostationary communication satellites, strategically placed in orbit above the equator, can act as mirrors reflecting such microwaves back to earth for reception at distant locations.

Cellular communication systems, as adjuncts to conventional telephone networks, have grown rapidly in popularity in the late 1980s. Cellular mobile radio telephone service, based on technical innovations achieved by AT&T, is a network radio design using low-power transmitters to serve small hexagonal areas ("cells") within a geographical region. Base station transceivers are placed in each cell and connected by wirelines to a central switching computer, and from there to the telephone system. The base station transceivers provide coverage for the mobile radiotelephones operating in the cell. Since the mobile radio signals may come from a vehicle near a cell boundary and spill over into the next cell, the frequencies used vary from cell to cell. AT&T proposed a seven-cell pattern of repeating frequencies, the configuration commonly used today. The availability of a wide block of the radio spectrum at 900mhz with appropriate propagation characteristics resulted in prompt approval of AT&T's proposal of the cellular concept by the FCC.

In the initial cellular system established in a particular market the cells can be quite large. As the market grows and the system approaches capacity, the cells can be made smaller and the same frequencies reused in nonadjacent cells, thus making efficient use of the radio spectrum.

A cellular mobile phone system installed in an automobile cost about $2,000 in the early 1980s, but by early 1990 such systems cost about $400. In addition, the customer paid about $100 per month for cellular service. Alternative service plans were being provided for high- and low-volume users. Cellular service sales in the United States were about $3.3 billion in 1989 and were expected to reach $4.4 billion in 1990. Cellular equipment sales were about $620 million in 1989 and were forecast to reach $655 million in 1990. The number of subscribers to cellular service, mostly doctors, lawyers, executives and other professionals, were estimated at 2.06 million in 1988 and 2.7 million in 1989. Cellular companies were planning on increasing capacities by introducing new digital technologies in the early 1990s, and were planning on marketing this new capacity to residential consumers at discounted prices.

By late 1989 miniaturized cellular telephones, with the same performance as larger mobile phones but small enough to be carried outside cars, were being marketed by Motorola and other manufacturers. While initial prices were in the $3,000 range, prices were expected to decline in the near future. As cellular telephone equipment becomes smaller and less expensive, the market is expected to continue its rapid growth. "We are looking at the dawn of an explosion in cellular telephone usage that will lead to stiff competition and better, more innovative products and services for consumers" (*New York Times*, January 28, 1990).

In 1989, cellular telephone systems relied on analog technology, and system congestion was becoming common during rush hours in urban areas such as Los Angeles, New York, San Francisco and Washington, D.C. Carriers began to plan the conversion of their cellular services to digital technology, which would provide three times the capacity at lower cost. The new digital equipment converts the sounds of speech into computer zeroes and ones which can be transmitted with less distortion than the more complex analog signals, while making better use of the spectrum and requiring less power. The first digital cellular networks are expected to go into operation in 1993, and will provide customers with facsimile services and access to data and information sources. At the beginning of 1990, the carriers with the largest numbers of subscribers were the Pactel Corporation, McCaw Cellular Communications Inc. and Bell South's Southwestern Bell cellular unit.

Personal communications networks are a still newer form of wireless telephone communication. These systems operate on higher frequencies than do those used by cellular systems and allow residential and business customers to carry lightweight wallet-sized phones everywhere. Personal communications networks have a shorter range than do cellular systems but also cost less to operate. The British government awarded licenses for personal communication networks to a number of international groups including Pacific Telesis Group, US West Inc., Cable and Wireless PLC and Millicom Inc. The equipment is expected to cost British customers about $300 each. New York–based Millicom Inc. has applied to the FCC for the use of a portion of the cellular radio spectrum so that a personal communications network can be offered in the United States in the early 1990s. Nynex and other local exchange companies are also planning tests of the wireless technology.

Teletext and Videotex services are a broad class of products providing electronic information services of various kinds. These services were invented overseas, and other countries, for example, Great Britain, are ahead in the establishment of commercial systems.

Teletext is a one-way service. Teletext information consists of characters, numerals and graphic images originating at a central computer keyboard or computer data base. The information is encoded, multiplexed onto a video signal and transmitted at a rate compatible with the color TV system. The information is contained on the unused lines in the vertical blanking interval of the video signal, and may be detected by a decoder attached to the TV set in the home. When the user punches particular keys on a keypad, the requested information is displayed on the screen, appearing either as an overlay or taking up the whole screen. The pages of information available are cycled periodically, and so the user must sometimes wait for the desired page to appear on the TV screen. Close-captioned TV programs are an example of this type of service.

Videotex is a two-way electronic information service. The information is generated and stored as in teletext systems, but the user requires a personal computer with keyboard and television monitor connected to the telephone system via a modem. A modem, or acoustic coupler, converts the computer's electrical output signals to sound signals that can be transmitted like speech over telephone lines. A modem at the receiving telephone converts the sound signals back into computer-compatible electrical signals.

The videotex system provides full two-way capability. The user may send or receive messages, accessing any part of the information available on computer data bases at any time. As current examples of videotex products in the United States, IBM and Sears, Roebuck, Inc. are principal backers of the Prodigy service mentioned in Chapter 4, and Bell Atlantic is offering the "Gateway" service. AT&T is planning on offering a national electronic directory of telephone subscribers.

Teletext and Videotex are not new inventions, but they are innovative new products in that they are being provided in very user-friendly forms and at very reasonable cost. In the early 1980s one had to be a computer expert to connect a personal computer to an compatible modem, which must in turn be connected to the telephone system. Then one had to learn the procedures for accessing various computer data bases with their different protocols. Most potential users are not comfortable with such complex do-it-yourself systems. The new services emphasize user-friendliness and versatility at low cost.

Distance is no longer a factor in the cost of transmitting information when satellite relay stations are employed, with profound implications for the structure of the long distance telecommunications industry. A communications satellite contains up to 30 transponders, each capable of receiving and rebroadcasting television, high-speed data or voice signals, both domestic and international. Signals may be frequency modulated or amplitude modulated, and the newest satellites are capable of providing 10,000 voice channels.

The newest technological frontier in telecommunications is optical electronics, often called photonics. Photonics components are devices that replace electrons with photons, and include optical-fiber cables, laser-based transmitters and receivers and devices for packing signals, switching them to different destinations and regenerating them. A single thin strand of glass fiber carrying pulses of laser light can transmit hundreds of thousands of times more information than can the traditional pair of copper wires. The obvious communications applications for fiber-optic technology involve the transmission of high volumes of information over long distances. Fiber-optic cables have been layed across the United States and across the Atlantic and Pacific oceans, and soon the earth will be circled by a fiber-optic network. In addition to efficient, high-speed transmission of voice and data over long distances, fiber-optic systems can carry video signals.

As the cost of fiber-optic systems declines, the regional telephone monopolies in the United States are experimenting with the replacement of conventional copper-wire residential telephone lines with fiber-optic networks. Such networks would greatly increase the number and types of services that could be efficiently provided by the local telephone companies. Such "fiber-to-the-home" networks will revolutionize residential telecommunications.

AT&T opened a new Solid State Technology Center near Allentown, Pennsylvania, in 1988. This center is intended to accelerate the process of designing, manufacturing and marketing new telecommunications products. "The old Bell system made everything it needed in a controlled way so it built the best system regardless of cost. Now our challenge is to come up with the best cost-to-performance ratio," said David Lang, a director at AT&T Bell Laboratories (*Scientific American*, October 1989, 74). A new emphasis on the whole innovation process rather than just on technology is clearly present at AT&T, the preeminent telecommunications company in the world.

In late 1989 both AT&T, Northern Telecom and other manufacturers were offering full fiber-to-the-home networks, and many local phone companies were negotiating contracts for or studying pilot projects. In fall 1990 the *New York Times* reported 21 separate experimental fiber-to-the-home experiments in the United States. Southern Bell had contracts with several manufacturers for evaluation purposes, and GTE was installing fiber-optic cables into 600 homes in Cerritos, California, an affluent suburb of Los Angeles.

The initial tests were concentrated in the kinds of affluent neighborhoods most likely to constitute the best markets for fiber-optic services, and this policy has generated complaints that the traditional policy of universal telephone service is being abandoned. The selective installation of fiber-optic networks may give both certain economic classes and certain geographical locations substantial economic, educational and health advantages. In addition, the ability of telephone companies to offer cable TV services over fiber-optic cables is generating a good deal of regulatory and congressional interest. Under 1984 legislation designed to protect cable TV companies, telephone companies are barred from delivering video signals to the home.

However, several bills permitting telephone companies to own cable TV franchises have been introduced. In late 1989 Senator Albert Gore, Jr. (D–Tenn.) and Representative Rick Boucher (D–Va.) were sponsoring legislation that would permit telephone companies to offer their own cable television services. The FCC, during the same period, announced that it was studying the feasibility of increased competition in cable television markets. The prospects for most local telephone companies eventually offering a wide variety of video and information services over fiber-optic cable networks in competition with cable television companies appear excellent.

Where AT&T has designed a local digital telephone system with a central office that acts as a hub for a collection of remote terminals that are in turn located near the homes served, the Raynet company, a small company working with Nynex, has designed a bus architecture. The Raynet bus runs through a residential area with several dozen links to curbside nodes, each of which is connected to three or four houses. Raynet's approach reduces the number of optoelectronic devices needed, as well as allowing the perhaps temporary substitution of high-speed copper wire for the short connection between curb and house, and thus reduces the cost per residence served.

In 1989 Nippon Telephone and Telegraph was planning to spend the equivalent of $200 billion on a national fiber-optic network in Japan. It has been estimated that a similar amount of money would be required to replace the copper-wire network that provides telephone service to the 90 million homes in the United States. A number of European countries are also interested in fiber-optic communication systems.

"The fiber-optic network becomes a pipe rather than a channel," says Norwood Long, Director of Fiber-Optics at AT&T's Bell Laboratories. "You can approach infinity in terms of what you can put over the pipe" (*New York Times*, November 5, 1989).

CHAPTER 9

Market Structure and Innovation

Technological change is often regarded as being exogenous to the economic system, influencing economic growth but not being itself subject to economic influence. This view of technological change is probably accepted more for reasons of convenience in constructing conventional economic models than for its validity.

Joseph Schumpeter was one of the first economists to study both the effects of technological change on the economic system, and the type of market structure most conducive to technological progress. He (1962) described a dynamic model of the capitalist economy in which the process of innovation, the ongoing introduction of new products into the marketplace, was the primary mechanism of competition. Schumpeter believed that the social benefits of dynamic economic efficiency, characterized by increasing industrial productivity and the competitive replacement of obsolete products by new products, are more important than the social benefits of the price competition usually emphasized in static economic theory.

Since only large firms possess the financial resources necessary for R&D programs and since such firms can take advantage of economies of scale and have greater ability to appropriate the profits resulting from their innovative efforts, Schumpeter believed that large firms were the most effective innovators and that the dynamic benefits of their innovative activities offset the decrease in economic efficiency due to their market power.

Schumpeter's views are ideologically charged because there is a traditional hostility to the power of large unregulated firms in the United

States, many of his critics claiming that innovation is promoted to a much greater extent by small firms in competitive markets. Even if innovative activity is greater in the large firms that dominate their markets, there is concern whether such innovative activity may have detrimental side effects on society or on the environment, or whether the loss of consumer sovereignty results in less desirable types of innovative activity. Galbraith (1967) gives an eloquent statement of the possibility of the large, modern firm imposing its will on society for technical reasons.

Technological advances most often occur in the R&D departments of large modern firms, but such advances also occur in universities and other, less formal settings. Technological advances increase the menu of technological possibilities, but it is the conscious choice of one technology among these possibilities, and the decision to make the necessary investments to introduce a new product or process based on the chosen technology into the market, that constitutes the innovative process.

Gomory (1990) distinguishes between the ladder paradigm and the cycle paradigm as descriptions of innovative activity: "A major scientific advance is at last achieved. Engineers refine it through successive stages until it can be embodied in a useful new device." This is the ladder paradigm, the traditional picture of innovation, occurring entirely within a single organization. The lesser-known cycle paradigm, in contrast, consists of planned generations of incremental improvements in related products, perhaps occurring simultaneously at a number of firms.

Gomory cites the invention of the telephone, the airplane, and the transistor as examples of the ladder paradigm, noting: "The ladder paradigm describes the startup of an industry or the launch of a product. The cycle paradigm provides the right approach to planning the evolution of an existing product." Gomory suggests that the cycle paradigm, when pursued with discipline, timing and attention to manufacturing requirements, brings improved products to market faster, providing the sponsoring firm a technological lead. Gomory's point is that the emphasis is often wrongly placed on directed R&D in the United States, with the expectation that such research will inevitably result in the prompt introduction of competitive products. "In cyclic development, the income from sales of the product funds both expenses and the R&D for the next product generation." Thus, the cycle paradigm funds and exploits success while shunning failure. Although the ladder paradigm is often the appropriate form of innovative activity, the cyclic process is vastly more important to an economy over the long run, and it should be well understood when developing policy. Market structure and the existence of government regulation affect the economic incentives that would normally motivate the appropriate innovative process.

Technological change is usually defined as an increase in the menu of existing techniques for producing goods and services or an increase in the number or types of products that may be produced. The selection of new productive techniques or new products or both from the enhanced menu of technological opportunities depends on the relative economic advantages of these decisions. Included in the term *economic advantages* may be some extremely sophisticated economic and political strategies that may be intended to strengthen the firm's long-term position in its market by deterring competition, rather than to just exploit some immediate cost or market advantage. As Coombs (1987, 10) states: "The specific activities of R&D and innovation and the more general activity of strategic planning are therefore attempts to reduce the uncertainty of the future, and to give the firm a greater degree of control over its development in a negotiated environment."

The difficulty in developing an economic theory of the R&D process is that R&D is normally undertaken by firms large enough to find the costs worthwhile. Such firms are often oligopolists, and the economics of decision making in these firms, which often emphasize negotiations rather than competitive activity, are more complex than the more predictable aggregate outcomes of decision making by firms in competitive markets.

An innovation may involve the production of a new product, a product innovation, or the use of a new type of productive process that lowers the cost of existing products, a process innovation. Both types of innovations may be classified as radical or incremental. It need not be the case that the firm whose R&D program was responsible for the technological change is also the firm implementing the innovation. In other words, except for the patent laws, which are not always airtight, the growing menu of technology is basically available, often through the technical journals and conferences, sometimes through reverse engineering, for all interested firms to choose from. Some may seize the opportunity to innovate and others may ignore the opportunity, or, because of financing problems or ineptitude, delay implementation.

In capitalist economies the innovation decision is, with some exceptions, left to the firm operating under market incentives. There are other aspects of innovation that are the proper concern of government. Opposing theories of innovation center on cause and effect. Some believe that "science push" is the primary source of innovation, and some believe in a "demand pull" theory. These two theories have different implications concerning the proper role of government. The antagonism that has developed between proponents of each of the theories centers on the ideological question whether technological advances are primarily serving the private interests of a few or are addressing the general needs of society. For example, food additives, insecticides and livestock feed supplements may reduce agricultural production costs, but are the resulting economic benefits flowing to a few, encouraged by the "push of science," offset by possible harm to the general public? The generation of detrimental side effects, or externalities, in the production of goods is usually considered justification for governmental intervention.

Another important issue concerning government is the perceived fairness with which an innovation is introduced into the marketplace. Is the early provision of the innovation to only those who can afford the cost of replacing the existing obsolete product fair when the innovation may confer great educational, professional or health benefits? The present controversy over the replacement of conventional copper-wire residential telephone lines with fiber-optic cable is an example of this issue. There are many other examples in the education and health fields. Will affluent households, being able to afford fiber-optic cable and thus gain access to a wide variety of informational services, have an even larger economic advantage over poor households? Is the introduction of fiber-optic cable to residences by large regional telephone companies, regulated by state commissions, fair to the existing cable TV companies, franchised by municipalities, who stand to lose their investments?

Government plays a role in the process of technological change through patent laws, regulatory policies, monetary and fiscal policy, its policy towards education and its science policy. Government may also play a role in the innovation process by granting a firm a regulated monopoly franchise to exploit an innovation, barring other firms from using the innovation in the same market, or by foreign trade, regulatory and antitrust policies. Because services such as telecommunications that are vital to all sectors of the economy are also affected by government policy, economic growth is also affected. But different government agencies often have conflicting objectives with respect to encouraging technological advances.

The problem of government policy concerning innovation may have different solutions depending on the type of political and economic system under consideration. Government policy to encourage innovation cannot contradict the general economic philosophy of the administration in power. Different economic systems are characterized by having different means for coordinating the plans and activities of economic units. The price mechanism is such a coordinating device in capitalist systems, and central planning is used in socialist economies. In a capitalist society the coordination of economic activities is carried out by many institutions including firms, the market and the state. Since technological change may originate in any or all of these institutions, it is not possible to deduce theories of the process from the study of any single one of these institutions. It is necessary to study in detail the interaction of the three types of institutions. This is the approach taken in the present study of the economics of innovation in the telecommunications industry.

While it is clear that firms, through R&D programs, influence the advance of technology, it is also true that a firm organized to exploit an important innovation often later finds itself reorganizing so as to manage advancing technology more efficiently. A market that appears to be a natural monopoly when exploiting one technology may lose this characteristic through technological advances, the new technologies having constant rather than increasing returns to scale. A government agency following one regulatory policy may change its policy radically as a result of technological change. All of these types of behavior have been observed in the history of telecommunications in the United States.

As an example of the relationship between basic science, R&D and innovation, the introduction of the loading coil into long distance voice transmission lines is described in some detail. This technological advance, occurring around 1900, made efficient long distance voice transmission possible and led directly to the formation of the national telephone monopoly in the United States.

The copper wire long used in telephone lines has electrical characteristics that tend to absorb much of the electrical energy produced by the voice transmitter, this effect becoming more pronounced as the length of the line is increased. The reduction in signal strength, or attenuation, depends on the frequency of the signal, the faster-traveling higher frequencies being absorbed more than lower frequencies, distorting the quality of the sound. The combined effects of attenuation and distortion effectively limited the range of early telephone lines to about 30 miles. The early telegraph systems, transmitting Morse code signals that were less subject to distortion, could be used over longer distances before encountering a similar limitation.

In the 1850s William Thomson, the preeminent British theoretician in electricity and magnetism of the time, became interested in the problem of the deterioration of electrical signals transmitted over long distance underwater cables. Thomson provided an accurate description of the problem, but it was Oliver Heaviside (1850–1925), a self-taught British scientist, who proposed a general theory of the transmission of electrical signals in 1876.

Oliver Heaviside left school in England in 1865 at age 16 and, teaching himself Morse code and the elements of electricity, obtained a job in a Danish Telegraph Company two years later. Heaviside gained experience as a telegraph operator and technician, and by 1871 he had returned to England where he was employed by the Great Northern Telegraph Company handling overseas traffic. Heaviside became interested in the problem of electrical signal attenuation over long distances, and he did a great deal of study on his own. Heaviside had educated himself in his spare time to such an extent that he published important papers on electricity in 1872 and 1873. In 1874, Heaviside quit his job to study electromagnetic theory on a full-time basis, and he was never employed again, living and working alone in a room provided by his family and gaining a reputation as an eccentric.

James Clerk Maxwell's *Treatise on Electricity and Magnetism* exerted a profound influence on Heaviside, who was to spend the rest of his life refining and simplifying the mathematics used in Maxwell's theory. Maxwell, who died in 1879, had predicted that "an oscillating electric field in space would generate a magnetic field oscillating at the same frequency, which in turn, would induce an electric field and so on. This electromagnetic wave would propagate at the speed of light" (Nahin 1990, 122–29).

During this period, William Preece was the technical expert at the General Post Office, which operated the British domestic telegraph service. Preece, through experience, had come to believe that the electrical property that limited the range of telegraph transmission was inductance, and he developed principles of telegraph line design intended to reduce inductance.

However, Heaviside had concluded on the basis of theoretical considerations that "high inductance is an advantageous electrical property for transmission lines" (Wasserman 1985, 9). The two men wrote opposing papers during the 1880s in a heated debate over the question of the effects of inductance on long distance signal transmission.

The problem was how to keep a high-frequency signal from being distorted as it passes through a circuit. Some distortion arises because of resistance, which converts some of the energy in the signal into heat. Distortion also occurs due to inductance, the brief storage of the signal in a magnetic field, and capacitance, the brief storage of a signal in an electric field. In the 1850s, it was believed that telegraph signal distortion over long distances was due primarily to resistance, which could be reduced by using heavier wire, at significantly greater cost, and capacitance, which was more difficult to control.

Preece developed a rule of thumb setting a practical upper limit on the product of resistance and capacitance in the line. Heaviside's contribution was an explanation of the relationship between capacitance and inductance in circuits, suggesting that increased inductance was needed to achieve a distortionless circuit over longer distances. With the correct inductance, the product of resistance and capacitance, which necessarily increased with distance, could be far larger than Preece allowed. "Heaviside imagined a device in the form of a tightly wound coil, called a solenoid, whose concentrated electromagnetism would greatly increase the circuit's inductance without adding much to its resistance" (Nahin 1990, 128).

Heaviside considered proposing that the General Post Office build such an inductance loading coil, but the idea was dropped because Heaviside's opponent, Preece, had a veto over research proposals in his organization. Some of Heaviside's papers referred sarcastically to the slow progress being made in long distance communications in Britain due to faulty knowledge, in comparison with the great strides then being made in the United States, where the British scientific work was well known. It was years before the British telegraph and telephone services, owned and operated by the government, benefited from the invention of the loading coil.

In 1874 Michael Pupin, age 15, arrived in the United States from Prague. While working to support himself, he took courses at Columbia University and, after receiving a degree, traveled to Europe for further study of the mathematical analysis of electricity at Cambridge and Berlin. He was in Europe during the period when the great advances were being made in electromagnetics. Returning to the United States, he was appointed to a professorship at Columbia.

Pupin was, of course, fully informed about the inductance controversy in the British scientific literature and, after completing his own investigations, patented a design for loading transmission lines in 1900. Pupin's patent application was filed at about the same time that George Campbell, a Bell engineer, was performing experiments on the use of loading coils on transmission lines for the Bell System. Pupin was apparently granted the patent over Campbell because his application contained a rigorous mathematical exposition.

By the late 1800s, the American Bell Company recognized the great economic significance of the range limitation in telephone communications. If this limitation could be overcome, a national telephone network would not only become possible, but also the Bell System would gain a great competitive advantage over the rival Western Union Company.

George Campbell was educated at the Massachussetts Institute of Technology, receiving a degree in civil engineering in 1891. He studied physics at Harvard during the next two years and then studied at Gottingen, Vienna and Paris on a fellowship. Campbell joined the Bell System in 1897. During the 1890s, Bell engineers had devoted much effort to the problem of inductance in transmission lines in order to increase the range of communications while making efficient use of material. Bell engineers initially believed that the use of bimetallic transmission lines, composed of iron and copper, could provide the necessary inductance. This approach appeared impractical to Campbell who, over a period of time, may have independently developed the concept of carefully designed and strategically placed loading coils along transmission lines (Wasserman 1985). Such "loaded" lines were not only effective at transmitting distortion-free signals over far greater distances, but less copper material was required, providing significant savings.

The Bell System purchased the rights to the Pupin patent for the large sum of $1 million, indicating the importance of this invention to the company. Note that possession of this patent, combined with Campbell's work, not only put Bell in a commanding position in the United States with respect to long distance telephone communications, but also closed the door to possible competitors in that market.

Pupin enjoyed a long and successful career at Columbia, where the physics building, still in existence, was named after him. Heaviside, who must be considered the originator of the concept of loading transmission lines, of course received nothing, and died in impoverished circumstances. Unknown to most students, Heaviside's simplification of the mathematics of Maxwell's electromagnetic theory is in common use in textbooks.

The story of the development of the loading coil provides important insights into the relationship between fundamental science, Maxwell's theory in this case; the recognition and description of the transmission problem by Thomson; the necessary idea for a practical application of the theory provided by the work of Heaviside; the rigorously developed and patented invention due to Pupin; and, finally, the extensive expenditures of time and money on organized experimental work performed by Bell engineers under the leadership of Campbell. All of these component activities were necessary in order to convert a scientific advance into an important commercial reality, and all took place apart from one another, driven by the common recognition of the problem.

This important innovation would not have occurred as early as it did in the history of telecommunications without the tradition of an international scientific literature open to all combined with the commercial incentives of a profit-oriented enterprise. The Bell System could now offer efficient long distance telephone service in competition with the existing telegraph service of Western Union, whose fortunes gradually declined thereafter. The introduction of long distance telephone service provided a strong incentive for independent local telephone companies to affiliate with the Bell System, which grew rapidly thereafter.

The ongoing search for more-efficient long distance telecommunications has never slackened. By 1990 AT&T had installed a trans-Atlantic fiber-optic cable, called TAT-8, at a cost of about $350 million. This cable is capable of carrying 40,000 phone conversations simultaneously at 280 million bits per second. The newer TAT-9 scheduled for introduction in 1991 has a capacity of 80,000 phone conversations at 560 million bits per second. In the summer of 1990 AT&T announced a new fiber-optic cable design with a capacity of 700,000 simultaneous conversations at 5 billion bits per second. It is interesting to note that the "cycle" paradigm seems to have been predominant over the years in the evolution of AT&T long distance signal transmission technology.

It is paradoxical that while large firms may be in the best financial position to foster the innovative process due to their market power, when such firms are regulated by government agencies they often lose the economic incentive to innovate. Innovation occurs in response to the prospects for above-average profit. Above-average profits are difficult to achieve and retain under effective government regulation, which is intended to eliminate them. In addition, the regulated firm is generally awarded an exclusive franchise and is therefore immune to the type of competitive threat that spurs innovation. No other firm has an incentive to engage in innovative activity in the same market. Regulated depreciation schedules may encourage or at least justify the continued use of obsolete capital equipment. When risky expenditures on new technologies are successful, they cannot be profitably exploited under the usual form of regulation. Such expenditures might not even be considered allowed expenses by the regulatory authority. The requirement to seek regulatory approval for new services can delay innovation for years. Finally, regulation, conforming to current political realities, tends to take a short-term perspective, while innovation is inherently a long-term investment process requiring a long-term perspective.

The question of the best market structure consistent with long-run economic efficiency, including the promotion of innovation at an optimal rate, is of fundamental importance, but of equal importance is the question of the impact of innovation on market structure. The recent history of the United States telecommunications market reveals the astonishing story of the disintegration of a 100-year-old regulated national monopoly, one of the most powerful companies in the world, because new technologies promising lower-cost services in certain specialized markets were exploited by several tiny competitors. The first competitive innovations permitted in the regulated telecommunications market set off a competitive chain reaction that is still continuing and expanding.

Innovation has the power to destroy monopoly market positions by changing cost structures, bringing about competitive behavior and re-quiring only a minimum of attention by regulatory authorities.

The driving force behind the dramatic shifts in the telecommunications busi-ness can be identified in a single word: technology. A plethora of innovations since the 1940s and 1950s have spurred the telecommunications industry, from a trickle of would-be competitors in the late 1960s to the wide-open equipment and long distance competition of today. And to the extent that bypassers are utilizing every kind of technology imaginable to avoid the local network, we can add local exchange service to that same competitive category. The net result has been just what one would expect in a competitive market: reduced costs and greater choices for the customer. (Danielsen and Kamerschen 1986, 87)

The central issue of the present work is the relationship between innovation and market structure in telecommunications, and the mix of the two that is likely to improve dynamic or long-run economic efficiency in the industry. This issue has already been settled in favor of competition in the interstate toll markets, and this decision looks better and better with the passage of time.

Local exchange telephone markets are still served by regulated mo-nopolies, ostensibly due to the existence of economies of scale in local telephone service. Is a region better served by a telecommunications industry dominated by a single regulated monopoly, established because it appeared to be most efficient in a static sense, with no competition and thus little incentive to introduce innovations? Or is society better served by a telecommunications industry with a few large firms in each regional market, under a minimum of regulation, each firm possessing the resources and economic incentives to pursue technological advances, but also possessing sufficient market power to take advantage of consumers? If such large firms are regulated by government, will they still retain incentives for carrying out research and development programs that ultimately benefit consumers? Perhaps society is best served by permitting free entry into all telecommunications markets, permitting any service to be provided by any firm on a "buyer beware" basis, no one firm having a sufficiently large market to require regulation? Will such firms have sufficient financial resources or sufficiently large markets to justify the expenses and risks of research and development programs? Will such a competitive market structure waste resources by needlessly duplicating expensive facilities? If a competitive market structure in telecommunications is encouraged, will it eventually collapse into a monopolistic structure due to fundamental economic factors, as in the case of the formation of the original Bell System? While this discussion attempts to address these questions with the hope of shedding some light, the answers may be unknowable.

The discussion first turns to the theoretical economics literature for guidance on the question of the effects of market structure on innovation. Arrow (1962) makes the argument that the economic incentives for cost-reducing innovation are greater in competitive markets because such markets are larger and the innovation may be patented by the developer and licensed to other firms in the same market. Arrow shows that the royalties due to the originator of a cost-reducing innovation in a competitive market are always greater than the value of the same cost-reducing innovation to a monopolistic firm. While Arrow's argument is logically correct in a world of perfect certainty, it may not be correct if the developer of the innovation cannot patent it and protect it from close imitations, or does not possess the financial resources to pursue patent violations in the courts. The thrust of Arrow's argument is that the more competitive a market is, the larger it is likely to be. Therefore, given the patent laws, competitive markets offer greater economic rewards to successful innovators.

A second argument against the claim that large firms are the most prolific innovators is that such firms are often regulated monopolies. As discussed above, regulated monopolies appear to have less incentive to reduce costs and, since they operate in a guaranteed market, are not under competitive pressure to introduce new products. An engineer assigned in New York City in 1900 to study the effects of competition in the telephone industry found that the Bell System usually responded to competition by improving service, and that the installation of automatic switching equipment by independent telephone companies provided

better service than that provided by the Bell companies at the same time
(Meyer 1980, 160). The slow depreciation of assets permitted in regulated utilities, together with an absence of competition, creates an
economic incentive to employ assets longer than would be economically
feasible in a competitive market. The higher costs, or equivalently the
lower productivity, of the aging assets are simply reflected in the tariffs
to the detriment of the public.

It is noteworthy that the local exchange telephone companies are all
state-regulated monopolies with historically low rates of technological
innovation. Over the years the Bell System has argued that its regulated,
vertically integrated monopoly, including its Bell Laboratories subsidiary, has been responsible for great technological change. It is true that
great technological advances have been made by the Bell System, but
many critics argue that these technological advances have not been
translated, promptly and generally, into improved local telephone service. Rather, it is the occasional periods of competition in the history of
the telecommunications industry that have been characterized by extensive telephone service growth and technological innovation. Finally, in
the years before the breakup, antitrust settlements had limited AT&T to
specified lines of business, slowing or deterring the conversion of
technological advances achieved by Bell Laboratories into innovations.

The form of government regulation may encourage or discourage
innovation. Baumol (1967) considered the effects on the rate of innovation
in regulated monopolies of different forms of regulation, concluding that
"regulatory lag" may provide adequate incentives for cost-saving innovations. Regulatory lag is the effect on the regulated firm of infrequent
regulatory action. During the period between rate adjustments, the firm
is free to retain any profits it may earn due to cost savings, these profits
perhaps providing sufficient incentives for innovation. The regulated firm
might have its allowed rate of return adjusted less frequently, permitting
it to retain interim cost savings due to innovations. This view that
less-vigorous government regulation can bring about improved economic
performance on the part of the regulated firm is becoming popular, but
many writers (e.g., Klevorick 1973) have disagreed with this sanguine
appraisal of the benefits of regulatory lag.

Meyer et al. (1980, 161) offer an argument in favor of more innovative
activity occurring in large firms. They point out that, compared with
small competitive firms, it may be easier for the large firm with market
power to appropriate all of the benefits of cost-reducing innovations.

In a paper with relevance to current developments in the telecommunications industry, Demsetz (1969) shows that if the monopolistic firm

can practice price discrimination, then the monopolist has a greater incentive to innovate. Monopolists will instinctively attempt to discriminate when setting prices for different classes of customers because it appears to be common sense. When the same product is sold in several markets, total economic gains are greatest when a higher price is charged in the less elastic market, the one in which the product is more highly valued.

Traditionally, regulation of telecommunications in the United States took the form of average cost pricing, which avoided both the relevant cost basis for pricing as well as price discrimination. Perhaps the effect of regulation on innovation in telecommunications can be improved if the regulated firms are permitted to discriminate between customer classes and between services to a greater extent than they have in the past.

In a related paper, Kamien and Schwartz (1970) suggest that the price elasticity of demand is important in determining whether or not a monopolist has a greater incentive to innovate than does a competitive firm. Kamien and Schwartz propose that government policies encouraging the development of new products by increasing price elasticity may be a more effective way to encourage innovation than is antitrust policy. Firms entering existing telecommunications markets by offering new products are often accused by the dominant firm of being imitators rather than innovators. The Kamien and Schwartz argument indicates that the products offered by the new entrants, by increasing the elasticity of demand as partial substitutes for existing services, also increase incentives for the dominant firm to respond by introducing innovations.

The present consensus in the theoretical economics literature seems to favor competition among a few large firms over either the regulated monopoly market or the classical competitive market as the best market structure to encourage innovation in a high-technology industry such as telecommunications. Large unregulated firms possessing both the sophistication and financial resources to conduct meaningful R&D programs and possessing adequate market power to exploit their innovations, yet not so large a factor in their markets as to ignore competition or attract vigorous attention from regulators or the Justice Department, may be the most efficient innovators. However, it is not at all clear that such markets are dynamically stable or sufficiently competitive to be self-regulating. We may confidently look forward to the emergence of new kinds of problems in the economic organization of the telecommunications industry.

The studies just cited are theoretical. A brief summary of empirical economic studies of the effects of market structure on innovation is provided. Most empirical studies of the relationship between market structure and innovation are concerned with unregulated industries rather than with the regulated components of the former Bell System. Economic statistics are analyzed in order to determine the extent to which innovation is affected by industry concentration, technological opportunities and barriers to entry. Proxy variables such as patent output, productivity increases and R&D expenditures are often employed to measure the rate of innovation. Industry concentration is assumed to be inversely related to competition, for which it is used as a proxy.

Stigler (1956), using data from the period 1899 to 1937, found that the highest productivity increases occurred in industries with falling concentrations, that is, increasing competitiveness. Phillips (1956), studying the behavior of different industries over essentially the same period, found, to the contrary, that productivity increased with increasing industry concentration. Meyer et al. (1980, 164) conclude that entry or a credible threat of entry promotes innovation. Therefore, the existence of barriers to entry reduces incentives for innovation. Scherer (1965), using patent output as a measure of the rate of innovation, found no statistically significant effect of industry concentration on innovation. As mentioned in a previous chapter, patent output often denotes defensive market strategy rather than innovative activity. When technological opportunity is included as an explanatory variable, contradictory estimates of the relationship between concentration and research effort have been found in various studies.

A case study commissioned by AT&T found that entrants into the telecommunications market increased product variety and introduced some innovations, but the effect of the entrants on the rate of innovation in the industry was unclear. An Arthur D. Little study, commissioned by AT&T, concluded that the effect of concentration on innovation is unsettled. The FCC, after reviewing a study it sponsored, concluded that entry resulted in AT&T exerting more effort to be responsive to customer requirements, but the study did not reach any general conclusions concerning the effects of market structure on the rate of innovation. Meyer et al. conclude only that "the empirical evidence thus far provides no support for the view that regulated monopoly is the best climate for innovation" (1980, 167).

Empirical studies of the effects of market structure on innovation may not properly treat the defensive use of technological change. A patent is not just an exclusive technological opportunity that may be transformed into an innovation. A patent is also a barrier to entry, and may be used to deter entry by potential future competitors in related markets. Competition among the various companies acquiring and holding patents on elements of radio technology at the turn of the century kept the companies from reaching cross-licensing agreements, and the commercial development of radio was delayed until well past World War I. The Bell System, one of the companies involved, may have benefited since its copper-wire telephone network remained free of competition from wireless communications for a number of years. It required action by the U.S. Navy in World War I to break the deadlock and expedite the development of radio technology during the war. After the war, the long-delayed cross-licensing agreements were completed, followed by the rapid commercialization of radio technology. Consequently, it may be a mistake to assume that the rate of innovation may be measured by the rate of patent grants or by expenditures on R&D.

The preceding discussion emphasized innovation and industry performance, a microeconomics question. The worldwide concern about economic growth and development after World War II resulted in growing interest in the relationship between advancing technology and macroeconomic performance. It was found that R&D expenditures did not automatically result in economic growth. The interaction between science, technological change and economic growth was far more complex than previously realized, and in the 1960s and 1970s industrialized governments were setting science policy by which funds could be channeled towards those technologies perceived as most related to the achievement of macroeconomic objectives. Beginning in the 1950s, detailed national studies of technological change and innovation processes began to appear, and distinctions between the various aspects of the processes were being drawn.

The United States relies on the competitive market system, constrained more or less tightly by a complex of government regulations and expenditures, to provide incentives for R&D activities. The competitive price mechanism requires several conditions to be fulfilled for its proper operation. Perfect competition requires many buyers and sellers, perfect knowledge, a homogeneous product and free entry and exit from the industry. In addition, it has recently been proposed that the effects of competition may still exist despite few sellers in the market if prospective new sellers are waiting in the wings for the appropriate opportunity to enter. This idea is contained in the new theory of contestable markets. In any case, the product produced in a competitive market is usually assumed to be a private good. The purchaser, by paying the market price, acquires title to the product for his exclusive use. A second consumer, paying the same price, obtains a different unit of the same product, and so on. When these conditions are fulfilled, competitive markets, in theory, create economic incentives for productive and consumption activities to take place at rates consistent with the maximization of economic welfare.

The competitive firm expending funds on R&D activities may not be able to appropriate all of the benefits arising from a technological opportunity. In spite of the existence of a patent, which may often be overcome by litigation, some other firm may be the first to convert the technological opportunity into an innovation, or imitators may quickly enter the market. Thus competitive firms may place a lower private value on R&D than that placed by society. In this case, too little R&D will be funded by such firms in competitive markets, and this is the usual rationale for the encouragement of R&D by government. However, government must be sophisticated enough to recognize the need for support of those R&D efforts where the beneficial externalities are the greatest, but there is general doubt that this sophistication exists, or if it does, whether it can withstand the inevitable political pressures of special interests. Large firms, on the other hand, can reap the economic benefits of their R&D activities and presumably have greater economic incentives to exploit technological advances, not requiring government action.

If only a few large firms exist in a market, collusive licensing agreements may have the effect of dividing the national market up between them, reducing the incentive for each to innovate even though the firms may possess a high level of technological expertise. The early development of radio in the United States was subject to this sort of delay. Patent rights and licensing arrangements are often used to delay or defend against innovation, and public policy must eventually address this problem.

Merely eliminating governmental regulation of an industry such as the telecommunications industry and permitting new companies to enter and compete with existing companies is not sufficient for society to obtain the benefits of a competitive market; the requirements of a competitive, or at least contestable, market must be fulfilled. The domestic airline industry provides an example. The industry, since deregulation, has promptly restructured itself so that most major air transportation markets are dominated by one or two carriers that have extensive price and quality-of-service discretion. In addition to the large capital investment needed, barriers to entry consist of the restricted gate capacities at many airports. Price discrimination, poor service and other monopolistic behavior is commonly observed and reported in the press. At the time of writing, the rate of innovations in air transport service is not at a historic high.

Professional economists emphasize economic efficiency, the overall performance of the economy, rather than distributional performance or equity. Governmental regulation of industry tends to emphasize equity, or at best equity and short-run efficiency. But in the short run the costs of lost efficiency gains through technological advances are small. Thus regulation, as a political activity, commonly protects the economic interests of existing producers and consumers at the expense of prospective new activities and products, to the detriment of innovation.

We may conclude that inventions and innovations are not exogenous to the economic system. Inventive and innovative activities are directed by allocations of funds to R&D programs as components of business strategy motivated by economic objectives, especially the objective of neutralizing competition. The most important activity of this sort apparently takes place in oligopolistic markets, and these markets are more difficult to understand, in the scientific sense, than competitive markets.

R&D generates new knowledge, new uncertainty, depends crucially on elapsed time, and purposely causes disequilibrium in markets. It is therefore difficult to reconcile R&D activity with the neoclassical assumptions of perfect information, movement towards equilibrium, and a common technology available to all. Innovations commonly involve the marketing of distinct new products, possibly produced alongside existing products or introduced in a sequential process, while the theory of the firm is most useful analyzing the continous production of a single commodity. Consequently, the conclusions of the theory of the firm under competition cannot be applied to the question of the effects of market structure on innovation without extensive modification. The discipline of game theory is much more useful in gaining an understanding of this issue.

An examination of the evidence in the case of telecommunications shows that large oligopolistic firms facing competitive entry are likely to be both more innovative than are either small competitive firms or monopolistic firms, and, as a result of constantly changing technology and contestability that provide more options to consumers, to have less market power than previously believed. Increased reliance on the discipline provided by competition and greatly reduced regulation, while occasionally subjecting customers to variable prices and inconvenience, is likely to provide incentives encouraging efficiency in the dynamic, or long-run, sense. While a few authors familiar with the telecommunications industry would disagree (e.g., Trebing in Nowotny et al. 1989), many more would agree (e.g., Wenders 1987, Meyer et al. 1980, and Danielsen and Kamerschen 1986).

CHAPTER 10 _____

Telecommunications Pricing and Innovation

For many years local telephone service has lagged behind long distance service with respect to the incorporation of technological advances. The costs of providing long distance telephone calls have fallen dramatically over time relative to the costs of providing local calls due to the uneven applications of advancing technology. Until the AT&T divestiture in 1984, federal and state regulators had been imposing a growing share of the total industry costs on revenues from the interstate market, thus heavily subsidizing local subscriber costs. The subsidization of local exchange operations as well as the economies of scale in conventional telephone service and regulatory restrictions on entry eliminated the possibility of unsubsidized competition in local markets, such competition often being the vehicle by which cost-reducing innovations are introduced. On the other hand, many observers were concerned about the possibility of encouraging uneconomic entry as an indirect effect of providing incentives for competition.

The focus of this chapter is on pricing and the rate of innovation in local exchange companies, since it is the local telecommunications markets that, through economies of scale in conventional technology, cross-subsidization and depreciation policies, were the most protected from competition and thus from pressures to introduce new technology.

The system of cross-subsidization of local exchange operations through separations and settlements could be made to work, however imperfectly, when telecommunications were a national monopoly, but such a system creates unacceptable distortions in an asymmetrically regulated telecommunications market. The same new technologies that reduced the long distance costs of the common carriers are also available to potential new entrants as well as to high-volume users of interstate services. High interstate toll charges can be bypassed by high-volume users who can build or lease private long distance capacity or purchase such services from a third party. Growing numbers of high-volume telecommunications customers have been doing exactly this during the 1980s, and defection by high-volume users has magnified the regulatory problem of raising sufficient revenues to cover local system costs.

It is now generally recognized that the introduction of competition into part of the national telecommunications network implied that competition would eventually increase throughout the system. Increasing competition, in turn, implies abandonment of the traditional regulatory philosophy of average-cost pricing in favor of prices approaching marginal cost in all telecommunications markets.

In 1970, AT&T, in response to the economics of changing technology, proposed a reduction in interstate toll charges, and this reduction was approved by the FCC. State regulators, concerned about the effects of this action on local tariffs, influenced Congress to reject the plan. The FCC consequently increased the local exchange subsidies being provided by interstate toll revenues rather than reducing interstate toll charges (Wenders 1987, 166).

In 1982 the FCC again attempted to reduce the distortions arising from toll subsidies, and again Congress opposed the plan (Docket 78–72). The FCC's plan recognized the distinction between usage-sensitive and non-usage-sensitive costs, proposing that tariffs providing revenues covering non-traffic-sensitive (NTS) costs should be collected from end users. The plan provided a mechanism by which NTS costs assigned to toll traffic by the separations process would eventually be collected from end users via a non-usage-sensitive charge. The essence of the FCC's plan is to return NTS access-line costs to the end users who cause them, putting the cost of interstate access on the access part of the customer's bill. Competition in the interstate market is inevitably impacting the regulated intrastate markets, and local exchange companies are under increasing pressure to move towards market-based rates. The access charge system proposed by the FCC was a move in this direction, but the inevitable political controversy has delayed its implementation.

According to Wenders (1987) cross-subsidization should be eliminated and telephone services unbundled where practical. Pricing for local exchange subscribers should include flat-rate charges reflecting the non-traffic-sensitive cost of providing local access and usage charges reflecting traffic-sensitive costs.

It is generally believed that flat-rate charges based on marginal cost will not provide revenues sufficient to cover the costs of providing local access because of economies of scale in local service. Therefore, if local access charges must rise above marginal cost, there is a strong case for charging different flat-rate local access charges to different classes of customers, those customers with the least elastic demands paying the highest charges. A system of prices whose relative values are based on inverse demand elasticities is called Ramsey (1927) pricing. Such a pricing scheme results in the smallest distortions in the quantities of the service demanded by the different classes of customers, and Ramsey prices are therefore considered "optimal departures from marginal cost pricing" (Baumol and Bradford 1970).

Wenders suggests that estimates of the magnitude of subscriber externalities, which provide a justification for setting the flat-rate charge for local access less than marginal cost, are approximately the same magnitude as the likely revenue shortfall due to marginal-cost pricing. Therefore, Wenders would set flat-rate access charges equal to the marginal costs of providing local access, with revenue shortfalls made up through usage charges for local calls.

Local usage charges are necessary to provide correct incentives for the consumption of conventional telephone services as well as for the introduction of innovative technologies and services. The usage cost of connecting and maintaining a local telephone call is very small. During the off-peak period, when the opportunity cost is also small and the costs of measuring the duration, time of day and distance of such calls may exceed the cost of the calls themselves, it may be appropriate to set the usage charge equal to zero. During peak periods, when by definition the system is used at capacity, the usage charge should be based on the costs associated with connecting and maintaining the call plus the opportunity cost at capacity.

Wenders discusses the sensitivity of the demand for local telephone subscriptions with respect to price. He notes that the demand for access is derived from the demand for usage, and that the typical residential customer's reservation price for access depends on the sum of the differences between the values the consumer places on the calls made per month and the price charged per call, plus the value of the calls received. In other words, the customer will subscribe if the access price is less than the sum of the net customer usage benefits per month. As the usage charge decreases, the rational customer will pay a higher access charge rather than do without service. Wenders quotes several studies showing that the demand for local telephone subscription is price inelastic, and that this inelasticity does not change significantly with income. The telephone in the home is practically a necessity in modern-day America, regardless of income. Wenders' point is that extensive subsidies to residential telephone service are not required to achieve universal telephone service.

Rohlfs (1979) provides data illustrating typical divergences of price from marginal cost in the Bell System in 1975:

Bell System Marginal Costs and Prices in 1975

Service	Marginal Cost	Price
Access	$12.50 per month	$10.83
Local Usage	$.035 per call	$0.00
Toll Usage	$.45 per call	$1.15

To summarize, all local exchange subscribers should pay a flat-rate charge equal to the marginal cost of providing access. Off-peak usage charges should be set at off-peak marginal cost, or zero in cases where these costs are very small and measurement is impractical. Peak-usage charges should be of block form, the first minute costing more than succeeding minutes to reflect the cost of gaining access, and remaining minutes based on marginal usage cost, including opportunity cost, during the peak period. If necessary to prevent mass defection from the peak time, shoulder peak pricing, also based on marginal cost, may be implemented (Wenders 1987, 107). Together, the flat-rate and usage charges should be set to provide sufficient revenues to support all local expenses, rising above marginal cost if necessary according to the Ramsey principle. Interstate access charges should cover the costs of providing access to the transmission facilities of the interstate common carriers, and interstate toll charges should be based on the marginal cost of such service, with peak-load pricing and cost-based volume discounts.

Not only will such a system of charges result in socially optimal consumption of telephone services and provide incentives for local exchange companies to make correct resource allocation decisions, but potential competitors, presumably employing new technologies, will be provided the unbiased economic information necessary when making decisions about market entry.

Until interstate toll charges are brought into line with marginal cost, there still remains the question of competitors taking advantage of the difference between toll charges and costs to enter the various segments of the interstate market. Such competition may not only be uneconomic in the interstate market, but may exacerbate the revenue problems of the local exchange companies.

Braeutigam (1979) calls such activity *intermodal competition*. Such entry may not be economic in the sense that it is needless duplication of the facilities of the common carriers, requiring higher than necessary prices while wasting resources. Einhorn (1987) suggests that the telephone utility should deter large users from choosing uneconomic bypass options by pricing high-volume usage below marginal cost. Low-level usage prices should exceed marginal cost. While this solution implies the subsidization of some high-volume use, charges satisfying these conditions can be designed so that each customer generates more revenue than cost. Einhorn emphasizes that when bypass is economically justified, then no system of prices can profitably or efficiently keep large high-volume users from choosing the bypass option.

Sharkey (1982) introduced the idea of "supporting prices" in a multiproduct firm. A firm charging supporting prices for a given production plan just breaks even, but any other production plan with smaller outputs, for example, that of a new entrant, loses money at the supporting prices.

The problem is to encourage cost-reducing or product-improving innovations through pricing incentives without encouraging uneconomic bypass of natural monopoly services. Ramsey prices are sufficiently above marginal cost to provide the requisite revenues, but they may encourage competition from entrants with costs higher than those of the incumbent firm. Such entrants can successfully compete by charging prices higher than the supporting price but below the Ramsey price for one or more of the products.

Brown and Heal (1987) propose that free entry into the industry be permitted, but that an excise tax on the product be imposed. The tax would equal the difference between the socially optimal Ramsey price and the optimal supporting price. If the potential entrant's technology permits charging a price at or below the supporting price of the incumbent while also paying the excise tax, consumers thus paying the same or a lower price than that charged by the incumbent firm, then competitive entry is economically beneficial.

Since local service was so heavily subsidized in the past, there was no possibility of economic gain if a new company were to enter that market.

As local rates move closer to cost, competition based on new technologies, such as digital cellular networks, is emerging in local markets, putting additional pressure on the local exchange companies to revise their pricing strategies and on state regulatory commissions to approve new pricing policies.

While technological advances accompanied by free entry and flexible pricing provide the primary economic incentives for innovation, the investment character of much innovative activity in telecommunications implies that investment incentives also influence innovation. All other things equal, a lower cost of capital will spur investments in facilities employing innovative technologies. Similarly, regulatory policies on depreciation affect the rate of innovation. The traditional regulatory philosophy permitted the slow depreciation of assets, which had the effect of increasing the base upon which the company's return was calculated, (Glaeser 1957). This policy implied using assets longer than they would be used in a competitive market in order to justify requested rates. As stated in an earlier chapter, AT&T wrote off $6.7 billion worth of obsolete transmission equipment in 1988 soon after the interstate toll market became competitive, incurring the first major earnings loss in 104 years. The AT&T writeoff occurred the same year that the FCC announced a new regulatory policy, "price caps," in which asset value no longer played a role in rate setting.

Local exchange companies, still regulated by state commissions, are beginning to complain about the adverse effects of slow depreciation schedules. The slow depreciation of existing assets was made feasible by cross-subsidization and the absence of competition, reducing incentives for innovations in local exchange services. As competition increases in local markets, local exchange companies now find themselves trapped by the slow depreciation schedules built into existing regulation, and their ability to respond to competition is weakened.

Most residential and many business subscribers pay a flat monthly charge for local telephone access but do not pay a usage charge for each local telephone call. Usage charges are necessary to provide incentives for the optimal provision of conventional local telephone service, new delivery technologies and innovative new services. Metering costs are declining and should not be an obstacle to usage charges in modern systems.

While there is a sound argument for regulating charges for access and local calls, since these are still provided under monopoly conditions, there appears to be no reason to regulate the introduction and pricing of innovative new products provided simultaneously over the same wires or delivered using new technologies. Such new products are subject to competitive market conditions, and their widespread availability increases economic welfare. The problem of commingling regulated and unregulated revenues is considered complex, but any common sense solution to this problem is socially beneficial in that more services are available to consumers on a voluntary exchange basis over essentially the same system and therefore at essentially the same access cost. Recent decisions by the FCC and Judge Greene appear to recognize the advantages of encouraging the local exchange companies to offer new products and services on a competitive basis by no longer requiring the companies to set up separate subsidiaries for this purpose.

Some writers believe that sufficient incentives for innovation in telephone service are already provided by the common form of regulation known as average-cost pricing, also called rate-of-return regulation. Averch and Johnson (1962) showed that rate-of-return regulation causes the regulated firm to perceive its after-regulation cost of assets to be less than the market cost, and thus such firms tend towards excessive use of capital assets relative to other inputs. Capron (1971) suggests that the Averch-Johnson effect creates incentives for the regulated firm to increase its use of capital investment in such a way that innovations are encouraged. The Averch-Johnson conclusion that the regulated firm's behavior is inefficient is based on the use of a static model that does not consider the question of dynamic efficiency. When criteria for dynamic, or long-run, economic efficiency are employed, it may be that the amount of capital investment encouraged by rate-of-return regulation is near optimal, according to Capron.

The implication of Capron's remarks is that the additional capital investment caused by regulation most likely involves the incorporation of innovative technologies into the production process and thus reduces costs in the long run, such effects not being considered by the static Averch-Johnson model.

As shown in previous chapters, close study of the regulated telecommunications industry in the United States certainly does show a preference for capital-intensive technologies, and the return to shareholders under regulation depends on such investment rather than on operating costs. But this preference on the part of the shareholders for capital assets does not necessarily promote adequate levels of innovation. As mentioned above, a large stock of capital assets can equally well be achieved by retaining economically obsolete but not fully depreciated assets in service longer than would be the case in a competitive market. The resulting higher operating costs are not a problem for a regulated utility immune from competition, since such costs can be passed on to customers in the form of higher rates, consumers having no alternatives. This is exactly the situation that competition can successfully exploit, and why permitting economic entry is necessary to encourage innovation. Socially optimal pricing forces the use of realistic depreciation schedules which, in turn, encourages prompt responses to competitive initiatives.

The positive Averch-Johnson effect on investment may be offset by the negative effect on the number of subscribers as access charges rise due to inefficient use of resources, and also by the lack of incentives for new-product introduction. Telecommunications executives are often strongly opposed to rate-of-return regulation because it tends to stifle innovation and discourage the introduction of new products (Weiss 1988).

There is another reason why the Averch-Johnson effect may not lead to increased innovative activity in local exchange companies. The effect is stronger to the extent that larger potential profits are possible, and the unregulated profit potential of local telephone service, lower than that of industry in general due to regulatory policies, may not have been sufficiently attractive to encourage adequate investment in risky new technology.

Significant economies of scale are widely believed to exist in conventional telephone local service, such belief supporting the continuing regulation of the Bell operating companies and the independent local exchange companies. Such economies may have encouraged local managers to take further advantage of existing technologies as expansion occurred rather than introduce new technologies that would not be competitive on a small scale and that could not be introduced on a large scale while significant undepreciated conventional assets remain in service. Consequently, regional managers may not have perceived significant cost savings through technological advances in the past, although this situation is certainly changing today under the actual or potential pressures of competition.

The possibility that the managers of local exchange companies might be more risk-averse than managers in general and therefore prefer to avoid investment in risky new technologies has been studied (Laughhunn et al. 1983), and there is no empirical evidence to support this view.

Assuming a normal attitude towards risk by telecommunications managers, there still remains the question of the attitude of the regulatory agency with respect to acceptance of the costs of technological initiatives as legitimate costs, including the provision of a risk premium in the regulated rate of return. In the past, risky decisions that were not successful were not always considered allowable costs by state regulatory commissions. Wenders (1987) emphasizes that price regulation should emulate price behavior in real markets, in which prices are often above costs as new products and technologies are introduced, and in fact it is the possibility of earning above-average profit that encourages competitive firms to invest in innovative activity. With luck, regulatory policy will become more sophisticated with respect to these issues. It is becoming recognized that permitting the regulated firm more earnings flexibility is in the long-run interest of the consumer.

In 1988 the FCC announced a "price cap" policy. Future regulated prices in its jurisdiction would be based on an average of current prices, corrected for inflation and for expected future productivity. If AT&T can achieve greater productivity than estimated, the company may retain any additional cost savings as earnings (Baumol and Willig 1989).

The discussion now turns to a theoretical model intended to illustrate the effects of pricing incentives on innovation in local telephone service. Following, for example, Mitchell (1978), it is assumed that the usage costs of providing local telephone calls are a linear function of the calls made, q. The usage cost per call is denoted by c_1, and the total annual usage costs are $c_1 q$. It is recognized that the usage cost of the first minute of a telephone call is greater than that of succeeding minutes, but this effect need not be considered in the present analysis, in which all calls are assumed to require the same length of time. There is no variation of demand with time. Time variations in demand for telephone service are addressed in the literature on peak-load pricing (e.g., Littlechild 1970), and the usual analysis is appropriate in the present application but is not explicitly included. Similarly, in the interest of clarity, it is assumed that only local telephone services are provided. The number of subscribers, a variable, is denoted N.

Traditionally, the local loop has been defined as the dedicated wire pair connecting the subscriber's equipment and the central office. Let the length of the average subscriber's loop in a market be L. The trend is to shorten the physical loop length, L, by moving switching equipment closer to the customer (Carne 1984). The total physical investment in customer lines is represented by NL. Digital technology now permits transmittal of more than one message at a time on a given line, and fiber-optic cable permits a range of new services, still only imperfectly defined, to be delivered simultaneously over the same line.

Therefore, for modeling purposes, it is assumed that technological advances in local telephone service may be represented by the number of customers served per unit of telephone line length, N/L, where L, the average length of a customer's line, may be effectively shortened in a given region through one or a combination of expenditures on new technologies. Larger values of the ratio N/L indicate that a higher level of the available technology has been selected for innovation. Many writers (e.g., Pujolle 1988) emphasize that the new technologies are reducing or eliminating the distance dimension of telecommunications costs.

Let the average access cost, including switching, per customer per unit of line length be c_2. Total direct customer or access costs are therefore c_2NL per unit of time under the assumption of constant returns to scale. If economies of scale in local exchange switching are included, local access costs become c_2N^aL, where the (positive) coefficient a is less than unity. At an average cost of capital of $r\%$ per year, total annual capital costs for subscriber lines and exchange switching are given by rc_2N^aL.

Suppose that, given the existing technological menu, the total levelized annual expenditure on innovation is proportional to the level of technology selected, N/L. Therefore, the total annual expenditure on innovation may be written as c_3N/L. It is assumed that this expenditure does not appear as an addition to the rate base, perhaps because innovative unregulated services are supported by this expenditure. But this annual expenditure lowers overall costs by reducing average line length, L.

It is assumed that customers for local telephone service have separate demands for local calls and for access that are represented, respectively, by the inverse demand functions:

$$P_q(q) \; ; \; P_N(N)$$

where p_q is the price charged per telephone call and p_N is the periodic access charge per customer. The usual restrictions on the signs of the derivatives apply. Mitchell (1978) provides a comprehensive study of the demand for local telephone service. Finally, the cost coefficients c_2 and c_3 reflect the provision of sufficient capacity to support total usage qN.

A welfare maximizing optimization problem is now formulated, the solution to this problem representing the solution achieved, in theory, by a competitive market. Forming an expression for total annual welfare W:

$$W = \int_0^q P_q(q) \, dq + \int_0^N P_N(N) \, dN - c_1q - rc_2N^aL - c_3N/L. \quad (1)$$

A welfare maximizing solution is found by setting partial derivatives of (1) with respect to q, N and L equal to zero:

$$\partial W/\partial\, q = p_q - c_1 = 0 \; ;$$

$$\partial W/\partial\, N = P_N - arc_2 N^{a-1}L - c_3/L = 0 \; ;$$

$$\partial W/\partial\, L = -rc_2 N^a + c_3 N/(L^2) = 0. \tag{2}$$

The above necessary conditions for a welfare maximum require setting the usage price, p_q, and access price, p_N, equal to respective marginal costs thus determining q^* and N^*, and choosing a cost minimizing level of technology, L^* where:

$$L^* = \sqrt{c_3 N^{1-a}/rc_2} \tag{3}$$

In equation (3), the socially optimal level of line length L^* increases with the cost of innovation, c_3. Line length decreases, that is, the level of technology selected increases, with increases in the cost of capital, r, and with the price of capital equipment, c_2. The total expenditure on innovation increases as N^* increases and as L^* decreases. These effects are all what would be expected in a competitive market.

In contrast to the above welfare maximizing solution, rate-of-return regulation with a flat charge for access and no charge for local calls is very common in regional telecommunications markets. How is such regulation likely to affect the level of technological investment in regional telecommunications firms? In the spirit of the Averch-Johnson model, let s denote the allowed annual rate of return on investment where, as usual, it is assumed that $s > r$. It is also assumed that, at a zero price, the demand for local calls is now fixed at q' per customer, the ' symbol denoting a particular value. The behavior of the firm attempting to maximize profit under regulatory restraint is represented by the solution to the Lagrangian problem:

$$L = p_N(N) \cdot N - c_1 q'N - rc_2 N^a L - c_3 N/L$$

$$-\lambda(p_N(N) \cdot N - c_1 q'N - sc_2 N^a L - c_3 N/L). \tag{4}$$

Setting partial derivatives equal to zero and rearranging:

$$MR_N = c_1 q' + c_3/L + r'ac_2 N^{a-1}L; \tag{5}$$

$$L' = \sqrt{c_3 N^{1-a}}/r'c_2. \tag{6}$$

where $r' = (r - \lambda s)/(1 - \lambda)$, the perceived cost of capital under regulation. When the regulatory constraint is binding on the firm, λ is a positive fraction and r' is less than the actual cost of capital, r.

Since MR_N rather than p_N appears in (5), and $c_1 q'$ is also included in (5), the number of subscribers provided access, N', is less than the optimal number, N^* from (2). However, when the regulatory constraint is active, that is, when market and cost conditions would permit a profit yielding a return on investment greater than the allowed return s, then the lagrange multiplier λ, $0 < \lambda < 1$, is positive, and the perceived cost of capital, $(r-\lambda s)/(1-\lambda)$, is less than r, which tends to increase investment in lines rather than technology, that is, increase L' in (6). The incentive to increase the investment in conventional lines beyond the welfare maximizing level L^* in (3) is an example of the well-known Averch-Johnson effect. Since the number of subscribers N' under rate-of-return regulation and flat-rate pricing is less than the welfare maximizing quantity N^*, because access price p_N^* is lower than p_N', and since L' is greater than L^*, then innovation expenditures $c_3 N'/L'$ are less than the socially optimal level.

Therefore, rate-of-return regulation without usage charges provides little incentive to the regional telecommunications firm to develop and offer innovative new products or implement cost-saving technical advances.

The philosophy of regulation was briefly discussed in Chapter 9 where it was stated that regulation often has objectives other than welfare maximization in the economic dimension. If a welfare maximizing pricing strategy is not feasible for implementation by the regulatory agency for one reason or another, possibly due to the large existing body of relevant law, then rate-of-return regulation with usage and access charges is preferable to such regulation with only access charges, with respect to the provision of incentives for investments in new technology and the development of new products. The issue of common-cost allocation, sure to arise as a diverse mix of products becomes available, is a relatively minor problem. The existence of a range of products provided over essentially the same physical system should alleviate rather than complicate the fixed-cost allocation problem in the sense that many solutions preferable to the existing situation are available.

A rational regulatory policy for regional telecommunications companies must encourage the introduction of innovative new products and services at competitive market prices alongside existing regulated services. Such a policy should include access charges and usage charges based on marginal cost for conventional services, with divergences from marginal cost according to the Ramsey principle probably not necessary. Metering costs are decreasing as new technology is introduced and should not be considered an impediment. Usage charges for local calls, while economically correct, also introduce the proper business perspectives to both subscribers and telecommunications managers. Such charges, free of regulation for new products when market conditions are appropriate, provide incentives for the development of innovative new products and cost-saving investments in new technologies.

CHAPTER 11

Telecommunications and Productivity

It is frequently claimed that advances in telecommunications technology are transforming the way we perform our daily tasks at work and at home. Applications of microprocessors, satellites, fiber-optic cables and video systems are providing fast, inexpensive access to a wide variety of information that enables us to enjoy a higher quality of life while being more productive in the workplace. While these claims are quite plausible, it has proven exceptionally difficult to back them up with hard evidence, especially claims concerning the productivity gains due to the new telecommunications and computer technologies. The reason probably has to do with the infrastructure nature of telecommunications. Like roads and water supply systems, telecommunications networks indirectly increase many dimensions of economic productivity, but the particular mechanisms are complex and hard to measure, and the beneficial effects may be significant in some sectors of the economy but insignificant in others. The improved quality and reliability of modern telecommunications services, especially when these services are provided at lower cost, are likely to be difficult to recognize when analyzing statistics. Therefore, overall summary data on the effects of telecommunications on economic productivity are usually misleading.

Most governments of industrialized nations believe that an efficient telecommunications network is an important factor in economic growth, and a cursory examination of the positive relationship that exists between telecommunications infrastructure and economic progress in most countries seems to support this relationship, at least in the macroeconomic sense. "A considerable number of studies of this kind have been carried out, generally supporting the conclusion that a high positive correlation exists between the development of telephone facilities and the level of overall economic development of a country" (Moss 1981, 76). However, there is a good deal of uncertainty concerning whether a causal effect of telecommunications innovations on productivity exists, or whether telephones are simply a preferred consumer good in the wealthier countries, telecommunications services being known to possess a high degree of income elasticity.

There is good reason to believe that advanced information technologies help shape business strategies that in turn lead to new forms of business organization that respond faster, are more closely adapted to changing market requirements, and are more efficient. But management must first gain an understanding of the social and psychological aspects of the new technologies in order to put them to best use in the workplace. Consequently, the path between a telecommunications innovation and the ultimate impact of the innovation on business efficiency is likely to be long and diffuse, and the effects of such innovations on productivity, being delayed as well as acting indirectly, are difficult to measure using conventional theoretical models and statistical techniques.

Tyler (1981) identifies four important trends that serve as a backdrop for a study of the relationship between telecommunications and economic productivity:

1. In spite of the many technological advances, the overall rate of growth of productivity in the national economy is slowing.
2. A growing proportion of economic activity is concerned with creating, processing and transferring information.
3. Electronic data processing and electronic transmission of information are growing faster than at any time in history.
4. The costs of electronic data processing and transmission are falling while the capabilities of the processing and transmission systems are growing rapidly.

An attempt to reconcile the apparent contradictions in these four trends is made in this chapter.

There exist few studies illuminating the effects of telecommunications on productivity because of the public good, or infrastructure, character of telecommunications services, because telecommunications advances tend to improve the quality dimension of business activities rather than the quantity dimension, and because the advances in this area are due to the introduction of new products and services, an area that is particularly difficult to study using the conventional tools of microeconomic theory. A telecommunications innovation should be regarded as a potential input into the production of goods and services and should be evaluated just like any other conventional input. The difficulty is that the new telecommunications technologies are not just substitutes for the old; they change what is being done as well as how it is done.

Microeconomic theory is the best available methodological framework with which to analyze the productivity of inputs in economic activity, but this theory is poorly suited to the analysis of telecommunications services as inputs for a number of reasons. Modern microeconomic theory is centered on the study of markets, and microeconomists employ abstract models of decision making in firms as an economical, operational way quickly to come to grips with the primary subject of interest, market behavior. But the abstractions so useful in the study of market behavior obscure the role of technological advances and innovation in firms. The well-known "ceteris paribus" assumption, interpreted as "other things being equal," effectively removes technological change from the usual comparative statics analysis. The neoclassical theory of production represents technological change as a movement of the production function towards the origin of the input axes, implying that fewer resources are necessary to produce the same quantity of output. The theory therefore reveals how the output of an existing production process can be increased through technological advances such as process innovations, but is silent concerning product innovations. "Only cost reducing improvements can be described by the production function. Improvements in performance or the appearance of new services find no place in the neoclassical theory of production" (Coombs 1987, 29). Although telecommunications advances have become available at lower cost, it is the dramatically superior and distinctly different capabilities of these advances that are considered important reasons for their rapid acceptance and justify increasing total expenditures on such services. But when modern telecommunications services simply shorten the time spent in contacting associates and clients, allowing more time for one's own paperwork, can it be shown that the office is more productive in any measurable sense?

A second source of difficulty when relying on the neoclassical theory is that it emphasizes static efficiency, penalizing those uses of resources that do not yield results in the present but that improve dynamic efficiency in the long run. Investments in telecommunication innovations undoubtedly have diffuse beneficial effects over long periods of time that may not be evident unless relevant data are carefully collected and analyzed. The superior performance of the economy in the dynamic sense due to the optimal incorporation of information technology is not likely to be apparent when using the static criteria of the neoclassical model. In fairness, it should be stated that modern statements of neoclassical theory include the general concept of decision making over time, but it is still true that the criteria for dynamic efficiency are rarely applied even in a qualitative sense.

A third reason for the difficulties encountered when using microeconomic theory in this application is that telecommunications services generate important externalities, beneficial effects extending far beyond the party purchasing the service. These externalities are difficult to measure and are often overlooked. Therefore, conventional productivity measures are likely to understate substantially the true productivity of modern telecommunications services.

A growing body of literature on managerial decisions and the organization of firms has been accumulating since the 1960s. Called the post-neoclassical theories of the firm, these theories are not as well developed as the neoclassical theory and, thus far, lack substantial empirical support. But these theories are probably more useful in assessing the impacts of telecommunications on productivity, although the assessment must be qualitative owing to the lack of pertinent studies and data. Several of the most prominent theories are briefly described here, indicating their likely relevance to the relationship between telecommunications and productivity.

Marris (1966) introduced sociological and psychological dimensions of managerial performance into the theory of the firm, emphasizing the importance of power, status and creativity as incentives for corporate performance, in addition to the traditional assumption that performance is determined by the compensation incentive. The probability of managers achieving the desired mix of economic and noneconomic goals is greatly improved when the firm achieves its economic objectives, which are likely to be broader than the simple profit maximization motive of neoclassical theory. In addition to psychological and economic incentives, managers are certainly concerned with stability and security both for themselves and for the organization, and this implies an emphasis on firm size as well as on profits. The desire to avoid uncertainty runs counter to the desire for economic success, since the market growth that increases firm size and brings economic gain also creates the need for additional managerial talent, usually untrained, and increases the risk of failure in the new markets. Frequent face-to-face communication re-solves much of this uncertainty through improved coordination of effort and increased confidence in other members of the organization, but such personal communication is time consuming, becoming unbearably so as the organization becomes very large.

Modern telecommunications services efficiently satisfy most of the need for such interpersonal communications in large firms, playing such an all-pervasive role in such organizations that they are often taken for granted, while also reducing the external sources of uncertainty by providing immediate access to current business data and intelligence.

Williamson (1975) proposed the transactions-cost theory of business organization. Williamson's thesis is that the boundaries of business organizations are formed by analysis of the costs of negotiating contracts with external suppliers versus the cost of bringing the supplier of the desired services into the organization. Transaction costs are a general concept, specifically including the costs of uncertainty regarding timely performance of the contract. Thus, when transactions costs dictate, the heirarchical form of coordination is substituted for the market form of coordination, the results of these decisions collectively determining the structure and boundaries of the organization in the interest of improved economic performance. To the extent that Williamson's theory has validity, the tendency for telecommunications innovations to reduce business transactions costs is likely to have profound effects on the size, functions and structure of large business organizations, probably in the direction of decentralization and reduction in size while dramatically improving the capability to respond quickly to changing market infor-mation.

Transactions costs have a similar effect on consumer shopping habits, the rise of the suburban shopping mall being due to ease of access by automobile and propinquity of a variety of stores at a single location. But shopping transactions costs can be reduced still further by shopping from catalogs over the telephone, and this new business has blossomed in recent years, accompanied by growing difficulties for the traditional department stores. Thus, the nature of marketing has been profoundly affected by telecommunications advances, and one has the feeling that this is only the beginning of a marketing revolution.

Penrose (1980) has offered the idea that a firm is an aggregation of resources, both physical and human, and that growth depends on the managerial services that are provided by the firm's management talent. Managers create growth by developing new business opportunities and also set a limit on growth because trained managerial services are scarce.

To the extent that telecommunications innovations make limited management resources more productive, the firm can grow faster and become more profitable.

Hay (1983) reviews the post-neoclassical literature and reports that while there is little empirical verification for the theories of Marris, Williamson, Penrose and others, he believes that the inherent plausibility of these theories will continue to influence thinking about the nature of the firm. It is quite likely that the post-neoclassical theories will provide guidance in future studies of the relationship between telecommunications and productivity.

Like other forms of infrastructure, telecommunications services are largely public goods in the sense that the benefits arising from the installation of a new communications device at a desk accrue to all of the members of the organization, not just to the occupant of the desk on which the device is placed. The beneficial externalities of telecommunication innovations in industry, difficult to measure, are likely to be far greater than the direct benefits to individual users. Some idea of the magnitude of the externalities involved can be gained by noting the numbers of persons employed in related industries.

Freeman (1989) states that more than half of all employment in the United States in the 1970s was related to the information occupations, including R&D, education, entertainment and data processing. AT&T, just before its breakup, was second only to the U.S. government in number of persons employed. Vast numbers of people are employed in information-related sectors, and telephones are universally present in homes and in accessible public locations.

The advantages of the new telecommunications technologies are easy to identify. They provide quick access to information and permit ease of handling, storing, and communicating information, all at rapidly decreasing cost. What are the mechanisms by which these new communication and information capabilities are likely to affect productivity in the economy in a conceptual sense?

Electronic messaging cuts interruptions, permitting managers larger blocks of time in which to study information, form plans and think through decisions. Access to off-site data processing facilities speeds and improves planning and decision making. Automatic call answering saves substantial time for both callers and call recipients. Electronic word processing speeds the preparation of written communications while dramatically improving their quality. Word processor interfaces with electronic transmission facilities and facsimile transmission save critical time transmitting information to distant decision points in a large organization. Teleconferencing saves time, energy and the expense of travel. Electronic publishing saves time while cutting the cost of the physical movement of large quantities of paper. Information services such as electronic library card catalogs, teletext and videotex permit decisions to be made on the best available information while saving time and effort.

Time is a critical dimension of competition. The firm that can consistently respond to changing market conditions or changing technology with new products and services ahead of its competitors will possess valuable economic advantages. Modern telecommunications services make fast response possible, and therefore such services are indispensable in a dynamic competitive economy.

At home, modern telecommunications permit pay-per-view television, information services available over the telephone through a personal computer and home shopping from catalogs or from information provided on television using credit cards. The coming fiber-optic and personal communications systems will provide a still greater variety of services.

Not all of the new information capabilities are beneficial. The potential for exposing telephone subscribers to invasions of privacy and various nuisances and fraudulent activity is a matter of growing concern, as is the potential for spurious or fraudulent calls to businesses. The ability to schedule airline seats more closely has improved air transportation efficiency, but travel is not as pleasant, and if a flight is missed it is very likely that a traveler will have to spend an additional night away from home before obtaining a seat on a succeeding flight. Economic efficiency is not always gracious and pleasant.

Telecommunications innovations have uneven impacts on different economic sectors of the economy. First, these new capabilities primarily benefit the managerial, marketing and service sectors of the economy, often improving quality rather than quantity of work, and doing little to improve manufacturing and agricultural productivity besides permitting marginally better planning and inventory control.

The new telecommunications and data processing innovations are often introduced into firms after only a casual study of requirements or with no study at all. Personal computers first began to appear in offices without benefit of management planning and often in spite of management's disapproval because of the existence of central computing facilities. In many of these cases individuals actually used their own funds to make investments benefiting their firms. Studies of the decision process used in the acquisition of computer-assisted design equipment in construction firms indicate that rarely is any rigorous decision analysis of the type routinely used when planning manufacturing facilities performed. "Implementation of the new technologies in the form of innovations in business lags the development of new technologies due to institutional factors. . . . While investments in manufacturing facilities are sophisticated, investments in office technology are haphazard" (Moss 1981, xiii).

Management must become more sophisticated about the need to plan for and use the new telecommunications services to improve productivity. Because telecommunications and related innovations affect the entire organization due to extensive externalities, the planning of such systems should not occur on a department-by-department basis but should take place in a central office. While this commonsense approach is often followed, it is surprising to note the many instances in which it is not.

Often, when information systems have been properly planned, workers refuse to make use of the capability to share information for political reasons. Believing that "knowledge is power" and behaving as though the workers in the next office are the enemy, office workers frequently defeat the purposes of information systems. "Our knowledge of the technology of telecommunications is much greater than our knowledge of the social and economic factors that influence the use of the new systems" (Moss 1981, xiii).

The additional time that telecommunications innovations make available should be used to improve the speed and quality of work, not simply to serve as a safety margin permitting employees to take longer to accomplish the same tasks. Similarly, learning how to use all of the features of the constantly improving updates of word processing software, electronic networks and data processing systems may not be the best use of office time, and good managerial judgment must be used to control the unending flow of such updates. Workers often show a tendency to become so absorbed in the care and feeding of their new electronic devices that they tend to forget why they were hired. The costs of telecommunications innovations are always much greater than anticipated, as is the time required to implement such innovations. Electronic mail systems do not result in "paperless offices" but rather duplicate much of the previous interoffice mail traffic.

As the economy becomes more and more service oriented, the slow rate of productivity increase in the service sector can become a critical problem for the economy due to inflationary and other factors, and the existence of a telecommunications technology capable of improving productivity will be of little use unless it can be properly implemented and managed.

A much more subtle aspect of the new telecommunications capabilities is that the costly distance dimension of economic activity is practically eliminated in many instances, with profound consequences for industrial organization and market structure. When satellite transmission is used to communicate data, "it makes no difference to costs whether you are transmitting for five hundred miles or five thousand miles. The message goes from the earth station up twenty-two thousand three hundred miles to the satellite and down again twenty-two thousand three hundred miles. It makes no difference whether the two points on earth are close together or far apart" (Moss 1981, 162). Microwave and fiber-optic transmission costs are also becoming divorced from distance as digital technology is introduced and will soon become competitive with satellite costs.

The elimination of distance as a factor in the cost of information-intensive business reduces the economic incentive for maintaining large, centralized organizations, also tending to change market structures. Economies of scale in information transmission, processing and storage grow smaller with electronic advances. There is no point in gathering and processing information in central locations when the cost of accessing information is independent of its location. The trend is clearly toward distributed data processing, storage and dissemination. But, as stated above, efficient management of the new business arrangements is necessary if the cost savings available through the elimination of the distance factor are to be secured.

The dominant costs in information processing are labor costs, and these costs are going up sharply relative to hardware costs (Moss 1981, 164). Since labor costs vary greatly between countries, some nations will have a comparative advantage in information processing. For example, Taiwan and Korea are emerging as international centers of software preparation and coding. The business press reported in early 1991 that computer programmers in the United States, heretofore considered hourly workers and therefore entitled to overtime compensation, are being reclassified as exempt employees in many firms, presumably to save on labor costs. The diverging costs of the different inputs used in information processing and telecommunications are another reason for the difficulty in identifying the productivity increases due to telecommunications advances.

Capital investment per worker has been higher in manufacturing and agriculture than in the service sector. Capital investments in the service sector, in telecommunications and information services facilities, have not been planned as carefully as they have been in the manufacturing sector, and often not managed at all but left to develop on their own. As the service sector continues to assume a more and more important role in the economy, lagging productivity increases in services compounded by the relative cost increases of labor cited above can become a critical problem for the economy.

In conclusion, the economic benefits of innovations in telecommunications services are diffuse and difficult to measure, and they may even be lost through careless management. When well managed, the new electronic information and communication services become significant advantages in achieving success in a competitive world. In the home, the new telecommunications services will increase the existing menu of information, education and entertainment possibilities, thus enriching lives while expanding the marketing revolution that has already begun.

CHAPTER 12 _____

Trends in Telecommunications

The rapid advances in telecommunications technology that have occurred since World War II along with a regulatory policy that ignored economic logic resulted in economic pressures that eventually forced the breakup of the national telephone monopoly in 1984. Telecommunications markets in the United States are artifacts of government regulation, and regulation is shaped by the evolving body of law. The relevant law is based on the history of the resolution of economic disputes that, in turn, often arise in the course of business firms responding to technological opportunities. Therefore, responsive changes in telecommunications markets lag years behind the continually unfolding technological advances that cause the changes.

Almost all observers believe that future telecommunications markets in the United States will be more competitive and less subject to regulation at state and federal levels, but government regulation will continue to play a key role in many areas, often taking new forms. This chapter reviews the effects that present technological trends are likely to have on the future cost and demand characteristics of telecommunications products and services, and consequently on the form future regulation is likely to take.

Regulation will continue in situations where the potential abuse of market power by monopolistic firms is evident or the public interest demands it, but the experience of the recent past has educated everyone to the necessity for forward-looking regulatory policies that encourage economic efficiency and technological progress rather than static policies that merely protect the current economic positions of shareholders and consumers. We have learned that judiciously permitting entry into regulated markets can encourage beneficial innovations while reducing market power, using the discipline of competitive pricing to accomplish what has proven so difficult for regulators to achieve over the years. We have also have learned that very great sums of money can be lost when entry is not successful and firms are forced to leave the industry. Entry should not be encouraged, as it has been in the past when temporary economic opportunities were created by poor regulatory policies, unless such entry is on an economically sound basis.

The main argument usually cited for the continued government regulation of telecommunications is that AT&T and the Bell operating companies so dominate their markets that no competitor or group of competitors can hope to gain a large enough market share to reduce the market power of the established firms. While this argument is consistent with accepted economic theory, examination of the events surrounding the entry of MCI and US Sprint into the interstate exchange market shows that even a small degree of competition in a specialized market may be sufficient to provide new price information to both consumers and regulators, and that the widespread availability of this information may deter abuses of market power elsewhere in the industry by attracting regulatory interest and creating consumer price resistance. Also, manufacturing cost has been decreasing with advancing electronic technology in such a way that the abuse of market power can cause a badly priced new product to miss the market entirely, being supplanted by the next version or imitation. Therefore, a small amount of competition can be surprisingly effective in reducing market power if price information is made freely available and if innovations may be freely introduced into the market.

As telecommunication systems move away from land-line networks with high sunk costs and come to rely more and more on radio technologies, the natural monopoly argument against competition will continue to lose force because the new cost relationships support competitive industry structures. Microwave transmission not only created new entry possibilities but also proved that economies of scale in radio-based telecommunication systems become insignificant at low levels of output relative to market size. In the future, increasingly important regulatory issues will center on the allocation of radio frequencies to competing firms, and therefore the FCC will come to play a larger role in the regulation of local communications.

Historically, the FCC has allocated frequencies on some arbitrary basis such as "first-come–first-served." The economic values of the property rights of frequency allocations must be taken into consideration in the regulatory process if economic efficiency is to be improved (Diamond et al. 1983). Many informed writers believe that owners of the property rights to frequencies should be allowed to sell them just as any other form of private property may be sold, and that a competitive market in frequency property rights would ensure the best uses of these scarce resources while being consistent with the principle of freedom of information. A mitigating factor is that digital technology and precision-tuned radio equipment are permitting far more efficient use of the frequency spectrum, thereby greatly increasing system capacities and making possible entirely new markets for radio communications.

A prominent feature of the hard-wire telephone system is privacy. It is against the law to tap a telephone line unless with the approval of the court. As the complete privacy afforded by hard-wire communications is replaced by radio signals that are intelligible to anyone with the proper equipment, questions dealing with the role of the government in regulating material transmitted over the airwaves must be addressed as well as questions of privacy and freedom of speech.

Since technological change will continue to create new economic opportunities that in turn will provide pressures for changes in regulation and market structure, a brief summary of current technological trends relevant to telecommunications is given.

At the heart of modern telecommunications facilities are large-scale integrated electronic circuits whose capabilities have increased exponentially in the last decade, and promise to continue increasing in the future as their costs decline. A few years ago memory chips could store perhaps 16k or 64k bits of information. Today, chips capable of storing 256k bits are in common use, and chips that can store 1,000k are being introduced into the market. Often overlooked is the higher and higher processing speeds that can be achieved in equipment using these chips. Memory chips with several times the capability of present chips are in the planning stage. Powerful, reliable, inexpensive memory chips combined with equally capable microprocessor chips are bringing about revolutionary changes in both consumer products and industrial production processes. In addition to the electronic switching that takes place at telephone exchanges, microprocessors and memory chips are used in "intelligent" telephones and modems and in PBX equipment. Digital signal processors are microprocessors used to analyze and manipulate voice signals so that they can be compressed and transmitted more economically than can analog signals. Digital signal processors can also be used to recognize speech, and research on applications is proceeding rapidly. The price-performance ratio of electronic circuits has been declining at a rate of 20% per year for several decades, and this trend appears likely to continue. +STOP

In the United States today, 80% of the common carriers' local loops are capable of handling digital signals. Thirteen percent of the local loops connect to digital switches in the telephone company's central offices. Eighty percent of the links connecting the local telephone companies with the long distance carriers are using digital transmission, and 80% of the long distance switching centers have digital equipment installed (Rowe 1988). The long-haul links connecting long distance switching centers use mostly analog microwave technology because of the lower bandwidth provided by microwave. The early promise of satellite voice communications is in doubt because of the unacceptable time delays in satellite transmission of signals, although this problem may be solved in the near future. Satellites are perfectly suitable for the transmission of broadcast information where the slight time delay is not noticeable.

The widespread availability of enormously capable yet inexpensive integrated electronic circuits combined with the decreasing cost of fiber-optic transmission has permitted the digitalization of communications networks, and the process of changing over from traditional analog transmission to digital transmission has already begun. "Optical-fiber technology is making the biggest impact on communications circuits today. Common carriers, as well as many private companies, are installing fiber cables at an unprecedented rate" (Rowe 1988, 472). The rate of transmission of data over fiber-optic cable is increasing dramatically, from 45 million bits per second today to well over 135 million bits per second in the newest systems coming into use. Increases in transmission capacity have direct economic impacts on both the cost of service and on sunk costs, the latter an important factor determining the extent to which the market is contestable and competitive entry is feasible. The distance between repeaters in long distance fiber-optic transmission systems has increased from 50 kilometers in the 1980s to over 100 kilometers by 1990.

The cost of fiber-optic cable is declining so fast that it will soon be economical to install fiber-optic cable and digital equipment to serve residential telephone subscribers. Test installations were being put into residential service in many areas of the country in 1990. Such installations are capable of providing a great variety of information services, and local exchange companies must be free to experiment with marketing and pricing policies in order to determine the best combinations of services to offer.

There is concern about the possible abandonment of the traditional policy of universal residential telephone service, since the economics of fiber-optic residential service may dictate installation only in higher-income homes. Lower-income families may fall further behind due to the lack of access to the rich information and educational opportunities offered over residential fiber-optic systems. This is a tricky regulatory issue because the usual approach, the subsidization of lower-income households by charging higher rates to higher-income households, might deter implementation of the new technology, and one must carefully assess the economic interests of those who have strong views on the subject.

The enormous increase in communications capacity attained through digitalization profoundly alters the cost structure of the industry that in turn is likely to alter the structure of the telecommunications market and eventually the form of government regulation. The regulated and unregulated divisions of AT&T, AT&T Communications and AT&T Information Systems, were permitted to rejoin and operate as a single business entity in 1986. The regulated and unregulated businesses of the Bell operating companies, which had been segregated into different subsidiaries, were also permitted to merge in 1986. The FCC now relies on cost-accounting standards to keep revenues from regulated and unregulated businesses separate. By the late 1980s the only barriers to entry being enforced were the traditional state restrictions on entry into the local exchange markets served by the BOCs and independent local exchange companies, and restrictions on entry into some intraLATA toll markets.

While much of the telecommunications market is open to competition and new companies have entered many sectors of the industry, the high cost of performing the R&D necessary to stay competitive will probably force most of the smaller entrants to merge or fail, leaving an oligopolistic market structure in the United States.

In Europe, only the United Kingdom is following the U.S. lead in encouraging competition, most other countries maintaining the traditional strict government control of telecommunications. In November 1990 the British government proposed opening the British telecommunications services and products markets, already the most liberal in Europe, to free competition. British Telecommunications PLC dominates the British market at present. Under a proposal announced by Trade and Industry Secretary Peter Lilley, U.S. and European companies would be free to introduce new telephone, mobile-radio and satellite services in Britain in the early 1990s.

It is expected that the seven regional Bell holding companies presently holding substantial shares of the British cable television market would be active participants. It is possible that the existing cable TV networks will be interconnected to form a competing telephone system in Britain, a market arrangement that would have extensive economic consequences to industry participants if permitted in the United States. Since 1984, when Britain's telephone monopoly was privatized, service has improved remarkably and inflation-adjusted prices have fallen by 20%. Other European countries are liberalizing their telecommunications policies, although basic telephone service is still usually provided by government-owned monopolies. Japan, with plans for a national fiber-optic cable telephone system costing several hundred billion dollars, is moving towards deregulation although lagging behind the United States and Great Britain. STOP

According to Trebing (Nowotny et al. 1989, 100), "If deregulated markets are to supplant economic regulation, they must be capable of inducing competitive behavior, diminishing market power, and compelling innovative performance. Achieving these objectives depends primarily on liberalized entry." Entry into the interexchange and interLATA markets has been substantial in recent years, and entrants continue to possess several advantages. They are not encumbered by high-cost, aging plants; they are assisted by the rapid growth of interexchange telecommunications markets such as MTS service, which has exceeded 12% annual growth since 1970; and they benefit from insignificant economies of scale in the new technologies permitting efficient operations at relatively small scale. Therefore, it can be argued that the interexchange markets are contestable and require only a minimum of regulation.

Local exchange service provided by the BOCs and the independent LECs is also vulnerable to competitive entry, according to the proponents of deregulation. While state regulatory commissions restrict entry to the franchised companies, by pass and diversion frequently occur, and private microwave systems, shared-tenant services and alternative-operator services are all increasing. The marketing of the formerly expensive cellular systems to residential users on attractive terms and the introduction of personal communications systems also constitute entry threats to the local exchange companies who, as noted below, are quickly moving into these areas to preempt competition.

Except for the CPE market, concentration is high in telecommunications markets. AT&T retains about 80% of the interstate switched voice market and a similar fraction of the private-line market. MCI has about 8% to 10%, US Sprint has about 4%, and all other OCCs share the remainder of the interstate market. The OCCs do not satisfy the usual economic criterion for contestability: reduction of the dominant firm's market share to less than 40%.

The competitive situation is worse in the intraLATA toll market which is dominated by the BOCs, these companies often carrying nearly 100% of the traffic in their areas. The intraLATA toll markets are usually protected from competition by state regulation.

AT&T controlled 95% of the overseas telephone revenues in 1986, with MCI and US Sprint sharing about 4%. Telex and telegraph, while valuable in certain specialized uses, are declining in importance and are not likely to constitute a competitive threat in the overseas market.

In spite of the great outcry over large customers bypassing the BOCs and the independent local exchange companies, these carriers have actually lost less than 6% of their business to bypass and similar alternatives. Entry of the "hit-and-run" contestable variety prevents incumbent firms from sustaining monopoly profits, from cross-subsidization, or from practicing price discrimination. Entry of the competitive variety causes the market share of the incumbents to decline with a consequent loss of market power and monopoly profits.

The theory of contestable markets concludes that a market with one or a few large sellers can be workably competitive if the existing technology is available to all firms, there are no sunk costs inhibiting entry and exit and consumers respond promptly to price differentials.

Trebing (Nowotny et al. 1989, 103) argues that the profits of the existing OCCs are not attractive enough to encourage additional entrants. He claims that sunk costs are monumental in the facilities-based carrier interexchange markets, citing the very large losses incurred by GTE and other firms as they left the business. He also cites the willingness of AT&T, the dominant firm in the interexchange market, to "strike back immediately with a variety of promotional pricing practices at any perceived threat to its market hegemony."

The local exchange companies do not appear to be contestable if conventional wire loops and switching facilities must be installed by competitors, since the large expenditures on lines and facilities would have no resale value if the entrant decided to leave the business. This argument depends on the continuing high-cost, restricted capacity and specialized marketing of cellular systems, all of which appear to be changing at the present time.

The most contestable telecommunications markets are clearly in enhanced services and the resale of services provided by facilities-based carriers. These businesses require little or no sunk costs, as access, transmission and switching services are provided by the common carriers. However, the resellers are at the mercy of the pricing practices of the facilities-based carriers, as restricted by the regulators, and so are increasingly forced to invest in their own facilities.

While Trebing cites many arguments against the possibility of competition surviving in the long run in telecommunications and therefore believes that regulation will continue to be an important factor in the market in the future, Danielsen and Kamerschen (1986) provide a very thorough study of the determinants of market power in telecommunications, and they are optimistic about competition being effective enough that only a minimum of regulation will be required, a view shared by this writer.

The wide range of potential new information service applications poses both problems and opportunities. While a variety of such applications promise to simplify our daily tasks and improve the quality of our lives, many applications also threaten our privacy, our need for peace and quiet, and our freedom from sensory bombardment.

The greatest change in telecommunications in coming years will be in the use of wireless technology for local telephone calls. Such systems, under study in Great Britain, will be as convenient as cellular telephones but cost much less. The prospect of intensive competition for cellular and conventional services has spurred research on wireless technology. "Wireless technology will be replacing the hardwire networks" comments Joel Gross, a telecommunications analysts for Donaldson, Lufkin and Jenrette. "The Bells' thinking is, if anyone will be bypassing their network, it should be themselves" (*Wall Street Journal*, December 19, 1990).

Mobile phone comanies are hoping to expand their markets beyond business and professional customers. A marketing survey conducted for Nynex concluded that for potential customers with incomes of $25,000 per year, one in every four, about 27 million in all, will probably buy a mobile phone in the next five years, and about 6 million plan to buy a mobile phone in the next 12 months. Cellular revenues have reached $3.2 billion per year, about $89 per month per customer. While many cellular industry executives still see the product as a business tool, too expensive for casual use by consumers, other companies perceive the profits that can be made by selective pricing and are offering deep discounts to casual users. Cellular companies are shifting their advertising from the business pages of newspapers to all media, selling equipment in department stores, electronics retailers and auto-supply shops.

In late 1990 there were indications that U.S. automobile manufacturers, impressed by the growing complementarity of telephones and automobiles, were planning on becoming major factors in the cellular communications industry. According to the *Wall Street Journal* (December 11, 1990): "General Motors Corp., eager to play a role in the cellular phone business, is gearing up to sell cellular phones and service through its dealers, and cellular network equipment through its Hughes Aircraft subsidiary."

A Hughes executive stated that GM was seeking a dominant position in the industry. Ford, Chrysler and other dealers were already selling and installing mobile phones in cars and acting as agents for cellular service companies. Hughes announced that GM was setting up a program for its dealers to sell cellular phone service over "The GM Network," an association of cellular service companies including Bell Atlantic, GTE and McCaw Cellular Commmunications.

Hughes plans to use a proprietary satellite technology, called digital speech interpolation, which will increase the capacity of future digital cellular networks by 10 to 15 times current capacity. Hughes will build and sell car phones capable of analog and digital transmission as well as providing the switching and radio transmission equipment composing the cellular network. Acknowledging that this is a very competitive business, Hughes representative Jack Shaw said, "But with the advent of new standards it kind of lets us play on a level playing field"(*Wall Street Journal*, December 11, 1990).

Air-to-ground telephone service was introduced by Airfone Inc., in 1984, and Airfone was acquired by GTE in 1986. In late 1990 the FCC approved the provision of air-to-ground telephone service by four additional firms: In-Flight Phone Corp., Clairtel Communications Group, Mobile Telecommunications Technologies Corp. and American Skycell Corp. In-Flight promised to be the most effective competitor, having spent $8 million developing a state-of-the-art technology and services including enhanced reception, facsimile capabilities and video games. In-Flight will utilize a digital aircraft telephone transmission system in contrast to Airfone's analog system. Airfone currently has phones installed in about 1,400 planes, about one-third of the U.S. domestic fleet (*Wall Street Journal*, December 21, 1990).

The paging industry, maturing after 35 years of operation, is feeling the growing competition from cellular phone firms. Paging firms are responding by offering new services such as voice mail and message transmission, and also seeking new markets and offering more convenient paging devices. According to the *Wall Street Journal* (November

20, 1990), "the outlook is so bleak that Nynex Corp. recently got out of the business, selling all its paging operations to Page America Group." Cellular phone prices are dropping so sharply that more and more people are finding it advantageous to be reached on a cellular phone rather than to be paged and then have to find a telephone to answer the call. The personal communications devices, sometimes called "midget cellular" phones, that are expected soon are likely to be so small and economical that paging devices will become obsolete.

AT&T and two regional Bell companies announced plans to develop a technology that could significantly increase the capacity of cellular radio systems (*Wall Street Journal*, August 3, 1990). AT&T signed an agreement with Qualcomm Inc., the company that pioneered the Code Division Multiple Access (CDMA) technology for use in cellular radio systems. Nynex and Ameritech plan to install the system in their networks, and if it is successful, the system is likely to become a standard for the cellular industry, replacing plans to move to the Time Division Multiple Access (TDMA) technology. These new digital cellular technologies are of enormous importance because they essentially free cellular systems from capacity limitations, and they promise profoundly to alter existing local telephone system cost relationships and market structures.

The intimate technological relationship that exists between electronic communications and computer technology is bringing these industries closer together. AT&T announced plans to acquire the computer manufacturer NCR Corp. for $6.12 billion in December 1990. Mergers of computer companies frequently fail, as did AT&T's previous joint effort with Olivetti, but AT&T spokesman McGinn said "In this instance we're talking about the same strategy, the same hardware, the same values" (*Philadelphia Inquirer*, December 9, 1990).

AT&T had accumulated $2 billion in losses in its computer operations over a six-year period, acknowledging that the link between the research being performed at Bell Labs and AT&T's computer operations was not strong, a defect that would be remedied by the acquisition of NCR. NCR executives were reported to be hostile to the planned acquisition, and at this time of writing it remains to be seen whether or not the offer is successful.

Further in the future, personal communications networks will allow millions of customers to carry lightweight pocket-sized phones, the so-called "midget cellular phones," by the end of the century. "We are launching the next generation digital cellular technology" said Allen Salmasi, vice-president of planning and development at Qualcomm.

"When PTS is introduced next year, a telephone number will no longer represent a place but rather a person" says Robert Keller, Nynex Mobile chief operating officer.

In July 1990, Bell South, in collaboration with the U.S. subsidiary of Sony Corp., announced plans to test certain aspects of the advanced wireless technology with as many as 300 customers in Georgia. Personal communications networks use pocketsize, cordless handsets that function much like cellular telephones, but cover smaller areas using low-powered base stations located closer together. The new wireless technology has the potential to be more lucrative than the cellular telephone business.

PCN America plans to construct experimental personal communications networks in Houston and Orlando. Millicom and Motorola Inc. also plan personal communications networks, Motorola proposing a satellite-based service (*Wall Street Journal*, October 10, 1990). In August 1990 Nynex announced plans to test a similar system in New York City. Nynex asked the FCC for an experimental license to test whether the new digital radio technologies could replace wire links between homes and telephone company locations. The test, scheduled to start in 1991, will be the first time a U.S. telephone company has tried to use digital radio transmission in its local loop to deliver conventional telephone service in metropolitan areas. Nynex said a wireless system could eventually improve service at a significant cost savings (*Wall Street Journal*, August 3, 1990).

Some telephone companies already provide radio links to remote farms, ranches and tiny mountain communities. Nynex said it plans to use radio transmission between New York Telephone facilities and the residential areas of New York City, Boston and White Plains, N.Y. Nynex announced that its Nynex Mobile Communications Unit plans to begin the nation's first wireless personal telephone service in Manhattan using the new CDMA technology. CDMA systems offer up to 20 times the call-handling capacity of current analog cellular systems by assigning an electronic code to each call signal, allowing more calls to occupy the same frequency space. TDMA, the currently accepted standard for future digital transmission, places each call in a time slot and transmits it within a single frequency. TDMA offers three to seven times the capacity of current analog cellular systems. In 1992, Nynex may introduce a portable wireless phone with a range of a block or so for neighborhood use.

The testing of local wireless communications systems is proceeding at other locations around the country. Bell Atlantic will be testing the new wireless technology in Philadelphia in the early 1990s. Pacific Telesis will test a system of its own when regulatory approval is obtained. U.S. West is cooperating with firms in England building a wireless system.

As one views the intense technological and marketing activity in the telecommunications industry at the beginning of the 1990s, it is difficult to believe that the industry is not strongly influenced by competitive pressures and that the deregulation of the interstate networks has not only been a success in that market but has also influenced the local exchange companies, who are behaving as though they believe that state regulation of local exchange operations will become more liberal as the new competing technologies become important in local markets. There are two reasons supporting this belief: the new technologies depend on radio frequencies, which have traditionally been regulated at the federal level rather than at the state level, and legal precedents permitting the introduction of new technologies do not protect the existing regulated local exchange carriers from competitive entry.

State regulation is gradually changing, reflecting the need to provide incentives for both innovation and efficient operations. In December 1990 New York regulators turned down Nynex Corp.'s request for a rate increase of $831.7 million, or 13%, plus depreciation charges of $190 million in connection with an equipment upgrade program. The New York Public Service Commission approved only a $250 million rate increase but increased the requested depreciation charges by $72 million. A financial analyst commented that "most companies would be delighted to write off their equipment faster." While disappointing to the company, the decision gave New York Telephone an added incentive to cut costs and operate more efficiently (*Wall Street Journal*, December 20, 1990).

As mentioned in an earlier chapter, in 1988 the FCC announced a new regulatory policy in its jurisdiction: "price caps." Such caps are based on current average prices and are adjusted in the future according to the rate of inflation and the expected rate of productivity increase. If the regulated firm exceeds the target rate of productivity increase, it may retain the cost savings as earnings, thus providing an incentive for dynamic efficiency. Costs saved through innovations benefit both the company and, through scheduled rate changes based on productivity increases, the customers. From an economic theory standpoint, such a policy may be socially optimal in the sense that no one is made worse off by shifting to such a policy and the potential benefits are great. The FCC applied price caps to AT&T in 1988 and to some operations of the local exchange companies in January of 1991.

Regulatory concerns do not just extend to the control of pricing and entry in the telephone business. The regulation of the broadcasting industry, including the regulation of content, are of great concern. The new cable television industry, composed essentially of monopolies in their service areas, offers the worst of both worlds: government control of broadcast information and monopoly. The rights to lay coaxial cables along city streets and utility poles is franchised by municipal governments. Once franchised, cable TV can become "the electronic reincarnation of the company store," according to Seattle mayor Charles Royer. Royer, chair of the National League of Cities Cable Television and Telecommunications Task Force, points out that there is only one source of delivery of programs and services. Competition by two or more cable systems existed in only 8 of 6,000 American cities. Cable TV served about one-fourth of the nation's households in the early 1980s, growing to about one-half by 1990.

In 1965 the FCC, under pressure from broadcasters, considered cable TV a service supplementary to conventional broadcasting, and rules were promulgated to avoid the audience fragmentation that would occur if too many programs were offered simultaneously. By 1979 the FCC reversed course and moved towards cable TV deregulation. In spite of many dire warnings, cable did not pose an economic threat to over-the-air broadcasters.

Federal regulation of telecommunications began with and continues to be governed by the act passed by Congress in 1934:

The Communications Act of 1934 subjected the telecommunications industry to a degree of central planning unprecedented in the United States. The recent trend towards deregulation reflects an almost universal belief that this experiment in central planning was a failure. Nevertheless, all attempts at reform, even those promulgated in the name of deregulation, have left the backbone of federal regulation untouched: centralized allocation of the frequency spectrum. (Diamond et al. 1983, 57)

The claim of public ownership of the airwaves has given rise to a centralized system of licensing that provides the legal and technical basis for many of the FCC's other rules and regulations. The cable industry depends on satellite and other relay services that use the spectrum for the distribution of programming. The FCC's initial regulation of cable TV was due to its interest in frequency allocation. The dramatic changes in telephone communications are due to new uses of the frequency spectrum rather than from laying new lines or building new switching exchanges. The issue is whether the frequency spectrum should continue to be treated as public property regulated by the FCC or whether private, freely transferable rights in radio communication should be created and a market for those rights introduced. A market would introduce the efficiencies of the price system into radio communications and permit entry that introduces more competition into the field (Diamond et al. 1983, 58).

Mark Fowler, FCC chair in the Reagan administration, called the Communications Act of 1934 "the last of the New Deal Dinosaurs" (Diamond et al. 1983, 5). The current outmoded regulatory environment developed in response to a market characterized by a certain degree of homogeneity of demand: person-to-person telephone calls, point-to-point voice or data communication, over-the-air radio and television broadcasting from stations within a given service area to receivers in that same area.

Today, demand is undergoing a fragmentation and specialization that has increased the number of players and produced a more competitive environment that was not foreseen when Congress determined the need to establish and protect natural monopolies in these areas. The Senate Commerce Committee described the 1934 act as "no longer adequate as a statement of national policy" (Diamond et al. 1983). What economists have called "technological exclusivity," the basis for exclusive franchises, no longer exists.

The regulatory framework in coming years is likely to be characterized by less regulation of business practices, managerial decisions and ownership, less centralization of regulation at the federal level, more-passive regulation in the form of benign oversight, and more exempt areas (Diamond et al. 1983, 18).

Government will retain authority over antitrust and market domination, copyright protection, spectrum management, technical specifications, fraud and consumer protection, privacy and information security, and health and safety.

Equally important, there must be an improvement in the public climate for telecommunications and a reduction in the uncertainty surrounding regulatory policy. FCC chair Fowler commented to a Senate committee:

To prepare for its key role in the coming information age, the telecommunications industry must attract billions of dollars of investment capital to continue its rapid pace of technological change. I believe that the necessary capital and the necessary innovation are most likely to be supplied in a competitive environment, free from government regulation wherever possible. (Diamond et al. 1983, 52)

The breakup of the Bell System in 1984 accelerated trends in information, computing and telecommunications that will continue to unfold in coming years to the great benefit of the public. Regulation of the industry must be eliminated where it no longer serves a useful purpose, and regulation must be altered so as to encourage economic efficiency in both the dynamic and the static sense where it is retained.

The dynamic benefits to society of competition in an industry subject to great technological advances are likely to be far greater than the cost advantages of regulated monopoly under the weak economies of scale that supposedly existed in the past, but that were probably lost in the costly bureaucracy of the former telecommunications monopoly and its accompanying regulatory agencies.

References

Arrow, K. J. "Economic Welfare and the Allocation of Resources for Invention." In *The Rate and Direction of Inventive Activity*. National Bureau of Economic Research. Princeton: Princeton University Press, 1962.

Averch, H. A., and L. L. Johnson. "Behavior of the Firm Under Regulatory Constraint," *American Economic Review* 52 (December 1962): 1053–69.

Baumol, W. J. "Reasonable Rules for Rate Regulation: Plausible Policies for an Imperfect World." In *Prices: Issues and Theory, Practice, and Public Policy*, edited by A. Phillips and O. Williamson. Philadelphia: University of Pennsylvania Press, 1967.

———. "On the Proper Tests for Natural Monopoly in a Multiproduct Industry." *American Economic Review* 67 (December 1977): 809–22.

Baumol, W. J., and D. Bradford. "Optimal Departures from Marginal Cost Pricing." *American Economic Review* 60 (June 1970): 265–83.

Baumol, W. J., J. Panzar, and R. D. Willig. *Contestable Markets and the Theory of Industrial Structure*. New York: Harcourt, Brace and Jovanovich, 1984.

Baumol, W. J., and R. D. Willing. "Price Caps: A Rational Means to Protect Telecommunications Consumers and Competition." *Review of Business* 10 (Spring 1989): 3–8.

Brandon, B. B. *The Effect of Demographics of Individual Households on their Telephone Usage in Chicago*. Cambridge: Ballinger, 1981.

Braeutigam, R. "Optimal Pricing with Intermodal Competition." *American Economic Review* 69, no. 1 (1979): 38–49.

Brock, G. W. *The Telecommunications Industry*. Cambridge: Harvard University Press, 1981.

Brock W. A., and D. S. Evans. "Predation: A Critique of the Government's Case in US v. AT&T." In Evans, D. S. *Breaking Up Bell*. New York: North-Holland, 1983.

Brown, D. J., and G. M. Heal. "Ramsey Pricing in Telecommunications Markets with Free Entry." In Crew, M. A. *Regulating Utilities in an Era of Deregulation*. New York: St. Martin's Press, 1987.

Brown, R. *Telecommunications: The Booming Technology*. Garden City, N.J.: Doubleday Science Series, 1970.

Capron, W. M. *Technological Change in Regulated Industries*. Washington, D.C.: Brookings, 1971.

Carne, E. B. *Modern Telecommunications*. New York: Plenum Press, 1984.

Coombs, R., P. Saviotti and V. Walsh. *Economics and Technological Change*. Totowa, N.J.: Rowman and Littlefield, 1987.

Crew, M. A., and P. R. Kleindorfer. *The Economics of Public Utility Regulation*. Cambridge: MIT Press, 1986.

Danielsen, A. L., and D. R. Kamerschen. *Telecommunications in the Post-Divestiture Era*. Lexington, KY.: D.C. Heath and Company, 1986.

Demsetz, H. "Why Regulate Utilities." *Journal of Law and Economics* 11 (April 1968): 55–65.

_____. "Information and Efficiency: Another Viewpoint." *Journal of Law and Economics* 1 (1969).

Diamond, E., N. Sandler and M. Mueller. *Telecommunications in Crisis*. Washington D.C.: Cato Institute, 1983.

Einhorn, M. A. "Optimality and Sustainability: Regulation and Intermodal Competition in Telecommunications." *Rand Journal of Economics* 18, no. 4 (Winter 1987): 550–63.

Evans, D. S. *Breaking Up Bell*. New York: Elsevier Science, 1983.

Faulhaber, G. R. "Cross-Subsidization: Pricing in Public Enterprises." *American Economic Review* 71 (December 1975): 966–77.

_____. *Telecommunications in Turmoil*. Cambridge: Ballinger, 1987.

Freeman, C. *The Economics of Industrial Innovation*. 2d ed. Cambridge: MIT Press, 1989.

Galbraith, J. K. *The New Industrial State*. Boston: Houghton Mifflin, 1967.

Glaeser, M. G. *Public Utilities and American Capitalism*. New York: Macmillan, 1957.

Goldberg, V. P. "Regulation and Administered Contracts." *Bell Journal of Economics* 7 (Autumn 1976): 426–48.

Hanley, P. A. "The Telecommunications Infrastructure Could Speed the Arrival of the Information Age." *Public Utilities Fortnightly*. (August 17, 1989): 22–26.

Hay, D. "Management and Economic Performance." In *Perspectives on Management*, edited by M. Earl. Oxford: Oxford University Press, 1983.

Henck, F. W., and B. Strassburg. *Slippery Slope: The Long Road to the Breakup of AT&T*. Westport, Conn.: Greenwood Press, 1988.

Kamien, M. I., and N. L. Schwartz. "Market Structure, Elasticity of Demand and Incentive to Invest." *Journal of Law and Economics* 13 (April 1970): 241-52.

Klevorick, A. K. "The Optimal Fair Rate of Return." *Bell Journal of Economics* 2 (Spring 1971): 122-53.

_____. "The Behavior of the Firm Subject to Stochastic Regulatory Review." *Bell Journal of Economics* 4 (Spring 1973): 57-88.

Laughhunn, D. J., R. L. Crum and J. W. Payne. "Risk Attitudes in the Telecommunications Industry." *Bell Journal of Economics* 14, no. 2 (1983): 517-21.

Leibenstein, H. "Allocative Efficiency versus X-Efficiency." *American Economic Review* 56 (June 1966): 392-415.

Littlechild, S.C. "Peak Load Pricing of Telephone Calls." *Bell Journal of Economics* 1, no. 2 (1970): 191-210.

Marris, R. *The Economic Theory of Managerial Capitalism*. London: Macmillan, 1966.

Meyer, J. R., R. W. Wilson, M. A. Baughcum, E. Burton and L. Caouette. *The Economics of Competition in Telecommunications Industry*. Cambridge: Oelgeschlager, Gunn & Hain, 1980.

Mitchell, B. M. "Optimal Pricing of Local Telephone Service." *American Economic Review* 68, no. 4 (1978): 517-38.

Moss, M. L., ed. *Telecommunications and Productivity*. Reading, Mass.: Addison-Wesley, 1981.

Nahin, P. J. "Oliver Heaviside." *Scientific American* (June 1990): 122-29.

Niskanen, W. A. *Bureaucracy and Representative Government*. New York: Aldine-Atherton, 1971.

Nowotny, K., D. B. Smith and H. M. Trebing. *Public Utility Regulation*. Boston: Kluwer Academic Publishers, 1989.

Owen, B. M., and Braeutigam, R. *The Regulation Game: Strategic Uses of the Administrative Process*. Cambridge: Ballinger, 1978.

Panzar, J. C., and R. D. Willig. "Free Entry and the Sustainability of Natural Monopoly." *Bell Journal of Economics* 8 (Spring 1977): 1-22.

Peltzman, S. "Towards a More General Theory of Regulation." *Journal of Law and Economics* 19, no 2. (August 1976): 211-40.

Penrose, E. *The Theory of the Growth of the Firm*. Oxford: Blackwell, 1980.

Perl, L. J. "The Residential Demand for Telephone Service." National Economic Research Associates (December 1983).

Phillips, A. "Concentration, Scale and Technological Change in Selected Manufacturing Industries 1899-1939." *Journal of Industrial Economics* (June 1956): 179-93.

Posner, R. A. "The Social Costs of Monopoly Regulation." *Journal of Political Economy*, 83 (August 1975): 807-27.

Pujolle, G., D, Seret, D. Dromard, and E. Horlait. *Integrated Digital Communications Networks*. Vol. 2. New York: John Wiley & Sons, 1988.

Ramsey, F. P. "A Contribution to the Theory of Taxation." *Economic Journal* 37 (March 1927): 47–61.

Rawls, J. *A Theory of Justice.* Cambridge: Harvard University Press, 1971.

Rohlfs, G. "Economically Efficient Bell System Prices." *Bell Laboratories Discussion Paper* 138 (1979).

Rowe, S. H., II. *Business Telecommunications.* New York: Macmillan, 1988.

Sankar, V. "Investment Behavior in the U.S. Telephone Industry." *Bell Journal of Economics* 4, no. 2 (1973): 665–78.

Scherer, F. M. "Firm Size, Market Structure, Opportunity and The Output of Patented Inventions." *American Economic Review* (December 1965): vol. LV, no. 5, 301–10.

Schumpeter, J. A. *Capitalism, Socialism and Democracy.* 3d ed. New York: Harper & Row, 1962.

Sharkey, W. W. *The Theory of Natural Monopoly.* Cambridge: Cambridge University Press, 1982.

Skinner, F. "Communications in the New Era." In A. L. Danielsen and D. R. Kamerschen. *Telecommunications in the Post-Divestiture Era.* Lexington, KY.: D.C. Heath, 1986.

Stigler, G. J. "Industrial Organization and Economic Progress." In *The State of the Social Sciences,* edited by L. D. White. Chicago: University of Chicago Press, 1956, 269–82.

_____. "The Theory of Economic Regulation." *Bell Journal of Economics* 2, no. 1 (Spring 1971): 3–21.

Tyler, M. "Telecommunications and Productivity: The Need and The Opportunity." In *Telecommunications and Productivity,* edited M. L. Moss. Reading, Mass: Addison-Wesley, 1981.

Wasserman, N. H. *From Invention to Innovation: Long Distance Telephone Transmission at the Turn of the Century.* Baltimore: Johns Hopkins Press, 1985.

Weil, R. L. "Allocating Joint Costs." *American Economic Review* (December 1968): vol. LVIII, no. 5 1342–45.

Weiss, W. L. *Ameritech: Annual Report 1987.* February 1988.

Wenders, J. T. *The Economics of Telecommunications.* Cambridge: Ballinger, 1987.

Williamson, O. E. *Markets and Heirarchies: Analysis and Antitrust Implications.* New York: Free Press, 1975.

Wright, P. *Spycatcher.* New York: Dell, 1987.

Zajac, E. E. *Fairness or Efficiency?* Ballinger: Cambridge, Mass, 1978.

Index

About the Author

JOHN R. McNAMARA is Professor of Economics at Lehigh University. After service as a naval aviator and experience in defense contract management, he took up his career interest in managerial economics. He has published articles in a wide variety of scholarly journals including *Operations Research*, *Naval Research Logistics*, *Managerial and Decision Economics*, *Water Resources Research* and *The Journal of Economics and Business*. He has also conducted sponsored research projects for government agencies and corporations.

THE ECONOMICS OF
INNOVATION
IN THE
TELECOMMUNICATIONS
INDUSTRY